The Unsettling Outdoors

RGS-IBG Book Series

For further information about the series and a full list of published and forthcoming titles please visit www.rgsbookseries.com

Published

The Unsettling Outdoors: Environmental Estrangement in Everyday Life
Russell Hitchings

Respatialising Finance: Power, Politics and Offshore Renminbi Market Making in London
Sarah Hall

Bodies, Affects, Politics: The Clash of Bodily Regimes
Steve Pile

Home SOS: Gender, Violence, and Survival in Crisis Ordinary Cambodia
Katherine Brickell

Geographies of Anticolonialism: Political Networks Across and Beyond South India, c. 1900–1930
Andrew Davies

Geopolitics and the Event: Rethinking Britain's Iraq War Through Art
Alan Ingram

On Shifting Foundations: State Rescaling, Policy Experimentation And Economic Restructuring In Post-1949 China
Kean Fan Lim

Global Asian City: Migration, Desire and the Politics of Encounter in 21st Century Seoul
Francis L. Collins

Transnational Geographies Of The Heart: Intimate Subjectivities In A Globalizing City
Katie Walsh

Cryptic Concrete: A Subterranean Journey Into Cold War Germany
Ian Klinke

Work-Life Advantage: Sustaining Regional Learning and Innovation
Al James

Pathological Lives: Disease, Space and Biopolitics
Steve Hinchliffe, Nick Bingham, John Allen and Simon Carter

Smoking Geographies: Space, Place and Tobacco
Ross Barnett, Graham Moon, Jamie Pearce, Lee Thompson and Liz Twigg

Rehearsing the State: The Political Practices of the Tibetan Government-in-Exile
Fiona McConnell

Nothing Personal? Geographies of Governing and Activism in the British Asylum System
Nick Gill

Articulations of Capital: Global Production Networks and Regional Transformations
John Pickles and Adrian Smith, with Robert Begg, Milan Buček, Poli Roukova and Rudolf Pástor

Metropolitan Preoccupations: The Spatial Politics of Squatting in Berlin
Alexander Vasudevan

Everyday Peace? Politics, Citizenship and Muslim Lives in India
Philippa Williams

Assembling Export Markets: The Making and Unmaking of Global Food Connections in West Africa
Stefan Ouma

Africa's Information Revolution: Technical Regimes and Production Networks in South Africa and Tanzania
James T. Murphy and Pádraig Carmody

Origination: The Geographies of Brands and Branding
Andy Pike

In the Nature of Landscape: Cultural Geography on the Norfolk Broads
David Matless

Geopolitics and Expertise: Knowledge and Authority in European Diplomacy
Merje Kuus

Everyday Moral Economies: Food, Politics and Scale in Cuba
Marisa Wilson

Material Politics: Disputes Along the Pipeline
Andrew Barry

Fashioning Globalisation: New Zealand Design, Working Women and the Cultural Economy
Maureen Molloy and Wendy Larner

Working Lives - Gender, Migration and Employment in Britain, 1945–2007
Linda McDowell

The Unsettling Outdoors

Environmental Estrangement in Everyday Life

Russell Hitchings

WILEY

This edition first published 2021

This Work is a co-publication between The Royal Geographical Society (with the Institute of British Geographers) and John Wiley & Sons Ltd.

Registered Office
John Wiley & Sons, Inc., 111 River Street, Hoboken, NJ 07030, USA
John Wiley & Sons Ltd, The Atrium, Southern Gate, Chichester, West Sussex, PO19 8SQ, UK

Editorial Office
9600 Garsington Road, Oxford, OX4 2DQ, UK

For details of our global editorial offices, customer services, and more information about Wiley products visit us at www.wiley.com.

Wiley also publishes its books in a variety of electronic formats and by print-on-demand. Some content that appears in standard print versions of this book may not be available in other formats.

Library of Congress Cataloging-in-Publication Data
Names: Hitchings, Russell, author.
Title: The unsettling outdoors : environmental estrangement in everyday
 life / Russell Hitchings.
Description: Hoboken, NJ : John Wiley & Sons, Inc., 2021. | Includes
 bibliographical references and index.
Identifiers: LCCN 2021015252 (print) | LCCN 2021015253 (ebook) | ISBN
 9781119549123 (hardback) | ISBN 9781119549154 (paperback) | ISBN
 9781119549161 (adobe pdf) | ISBN 9781119549130 (epub) | ISBN
 9781119549178 (ebook)
Subjects: LCSH: Human ecology.
Classification: LCC GF41 .H58 2021 (print) | LCC GF41 (ebook) | DDC
 304.2--dc23
LC record available at https://lccn.loc.gov/2021015252
LC ebook record available at https://lccn.loc.gov/2021015253

Cover image: © 4 PM production/Shutterstock
Cover design by: Wiley

Set in 10/12 Plantin Std by Integra Software Services, Pondicherry, India

Printed and bound by CPI Group (UK) Ltd, Croydon, CR0 4YY

C094140_070621

Contents

Series Editors' Preface

The RGS-IBG Book Series only publishes work of the highest international standing. Its emphasis is on distinctive new developments in human and physical geography, although it is also open to contributions from cognate disciplines whose interests overlap with those of geographers. The Series places strong emphasis on theoretically informed and empirically strong texts. Reflecting the vibrant and diverse theoretical and empirical agendas that characterize the contemporary discipline, contributions are expected to inform, challenge and stimulate the reader. Overall, the RGS-IBG Book Series seeks to promote scholarly publications that leave an intellectual mark and change the way readers think about particular issues, methods or theories.

For details on how to submit a proposal please visit:
www.rgsbookseries.com

Ruth Craggs, *King's College London, UK*
Chih Yuan Woon, *National University of Singapore*
RGS-IBG Book Series Editors

David Featherstone, *University of Glasgow, UK*
RGS-IBG Book Series Editor (2015–2019)

Acknowledgements

Whilst carrying out the four projects on which this book draws, I have benefitted from many supportive academic friends over the years. This started during my PhD research, which shapes Chapter 5. At that time, I shared an office with, and learnt a lot from, peers including Caroline Bressey, Jason Chilvers, Robert Doubleday, Tara Duncan, Andrew Harris, Elaine Ho, Jason Lim and Bronwyn Purvis. Gail Davies and James Kneale were consistently wise and encouraging supervisors too. Since then, a further series of colleagues, in particular Chris Bear, Rosie Day and Alan Latham, have taken turns in helping me to figure out the nature and scope of my academic work. Alan was also my collaborator on the project that shapes Chapter 4. Those who did their PhDs with me were pretty good at exploring the big picture too. Elizabeth Shove and Gordon Walker have been sources of inspiration from the north. Further afield, colleagues in Australia have always provided refreshingly honest exchanges about how we work whenever I came to visit (whilst also being very welcoming and a lot of fun). Two further Australians, Alison Browne and Tullia Jack (along with a rogue Yorkshireman), were part of cooking up the project that is discussed in Chapter 6. Alongside all these academic friends, many others (and more excellent research trips) have helped me to refine the ideas that I present in this book. Cecily Maller gave thoughtful comments on an earlier draft and so did Paul Harrison. The editors and reviewers for the RGS book series were a pleasure to work with too – striking a very welcome balance between encouragement and critical engagement. My thanks to all.

This book would have been impossible without the financial support that allowed me to undertake the four studies that provide it with a backbone of empirical evidence. The studies that are considered in Chapters 3 and 5 were funded by the UK's Economic and Social Research Council (RES-000-22-21-29; PTA-026-27-0465), the project discussed in Chapter 4 was supported by UCL's 'Bridging the Gaps' fund, and the field research on which I draw in Chapter 5 was supported by both UCL and the ESRC (RES-597-25-003). It is also definitely true to say that this book could not have been written without the help of all the different people who took part in these studies (for whom I should say upfront that pseudonyms have been used throughout). I enjoyed meeting them all and was frequently delighted by their willingness to talk with me about the detail of their everyday lives.

Alongside them, my family and friends have supported me in all sorts of ways. So thanks to you too!

Chapter One
A Wager and a Strategy

A Wager

If we want to understand the likelihood of future societies having regular beneficial contact with living greenspace, we should examine how outdoor experiences are handled by people in their everyday lives today. This is the first wager of this book. My suggestion is that, if we ignore how widespread social practices can serve to discourage people from a fuller engagement with the outdoors, a certain kind of environmental estrangement could become increasingly entrenched.

The Argument

This first chapter tells the story of how I came to make the above wager. It begins with some reasons for encouraging greenspace experience in everyday life. Then it considers why, despite the various benefits that have been linked to this experience, many people may be turning away from it. With that prospect in mind, I consider how a particular combination of concepts could shed a useful light on how this process is embodied. This chapter is therefore partly about existing studies of beneficial greenspace experience and how they handle the social trends that stand to shape the future of this experience. But it is also about how a particular

The Unsettling Outdoors: Environmental Estrangement in Everyday Life, First Edition. Russell Hitchings.
© 2021 Royal Geographical Society (with the Institute of British Geographers). Published 2021 by John Wiley & Sons Ltd.

set of ideas might help us to reconsider the challenges involved in tackling these trends. Here I am interested in how certain strategies for studying the relationship between humans and nature could be combined with a focus on how people are drawn into patterns of everyday living. The overall aim is to set the scene for a battle between the various apparent benefits of spending time with plants and trees and a series of commonplace social practices that could be separating people from them.

Greenspace as Home

Being near plants and trees appears to provide people with various benefits. One of the most arresting and influential studies to suggest this compared the recuperation rates of hospital patients with different views. The required information was already being collected by the hospital, but by looking at it with a fresh pair of eyes, Ulrich (1983) found that those patients who looked out onto areas of greenery recovered more quickly. Though this study couldn't tell us too much about the mechanism involved, clearly there was something about seeing living vegetation through the windows of their wards that helped some patients to get better sooner. Another well-known study suggested this experience can also benefit those who are not yet ill. Moore (1981) found that prisoners with cells facing internal courtyards use medical facilities more often than those overlooking fields further beyond. So, being able to see greenery may prevent health problems as well as speeding recovery once they have been medically addressed. We have also seen how, for residents of city estates, being able to see trees and grass from their apartment windows appears to help them handle the various challenges they are facing in their lives and even reduce aggression levels (Kuo and Sullivan 2001). Other field tests have shown how contemplating vegetation can reduce blood pressure (Van den Berg, Hartig, and Staats 2007) and improve mood and self-esteem (Pretty et al. 2005). A recent study to build on what is now a fairly well-established tradition of identifying and enumerating the benefits that greenspaces can bring to people suggests that spending time in these spaces can reduce the cravings of those who are trying to overcome various addictions (Martin et al. 2019). These are just a few examples (see Keniger et al. 2013, for many more). The point, however, is that, if we allow ourselves to see humanity as a collective whose members continue to share the same essential attributes, there is a lot of evidence for the benefits of being around greenspace.

Why is this? One of the leading arguments is that being near to living vegetation provides a valuable form of psychological restoration (Kaplan and Kaplan 1989; Kaplan, Kaplan and Ryan 1998). The suggestion here is that simply looking at greenery can help people to mentally recharge themselves since contemplating the intricacies of vegetation can temporarily beguile us in a manner that allows us to transcend our immediate worries before returning to our tasks refreshed (Kaplan 1993; Han 2009). Another possibility is that this experience naturally neutralises the stressed feelings that many of us may otherwise increasingly harbour (Ulrich et al. 1991). Some even work with the assumption of a fundamental connection between humans, plants and trees such that our history of co-existence instinctively inclines people to seek out the reassuring familiarity of environments that contain living vegetation. This leads directly to the 'biophilia' hypothesis (Kellert and Wilson 1993), understood as the innate attraction to natural processes that humans may possess. The contention here is that dwelling within, and profiting from, certain living landscapes was fundamental to our development as a species. We should therefore be unsurprised to observe a positive response from people today. For example, some have explored how this filters through into a preference for looking at particular species of tree and how, within that, the trees that helped us to prosper in earlier evolutionary times are those that we still most like to see (Summit and Sommer 1999). We could take this to mean that a desire for greenspace experience is hardwired into humans. Either way, and regardless of whether we buy into this idea or not, these studies, when taken as a whole, suggest that people can benefit in all sorts of ways from exposure to these environments, if they are given the chance.[1]

Tempting People into Parks

What should be done with this knowledge? If we now consider how societies have most often thought about the right response to these findings, a common next step is to turn to the provision and design of public parks and gardens. This makes sense. If most of us now live in cities, if researchers know that being in and around greenspaces can benefit people, and if one of the tasks of good government is to ensure the inhabitants of a planet whose humans live increasingly urban lives have access to the services that are good for them, then city parks and gardens become an obvious focus for policy. In line with this argument, a lot of effort has gone into thinking about the forms of park provision that stand to produce the maximum social benefit. In doing so, effective landscape design and urban planning has come to seem like the obvious means of putting these ideas into practice. Indeed, the path between studies of greenspace experience and suggestions about what should be done with their findings is now fairly well trodden. And it commonly moves from an argument about benefits to an interest in the most effective means of designing and planning the most visually attractive and welcoming city greenspaces.[2]

Recent examples include a study in which Chinese citizens were shown urban scenes (from those with lots of concrete to those with more vegetation) in an attempt to identify how public greenspaces could be most effectively designed to reduce stress (Huang et al. 2020). Then there is a consideration of the value of features like colourful flowers based on how people in British parks and gardens respond to different pictures of plants (Hoyle, Hitchmough, and Jorgensen 2017). Another example is an exploration of the extent to which 'actual' or 'perceived' biodiversity in the greenspaces experienced by French residents impacts most positively on their wellbeing (Meyer-Grandbastien et al. 2020). A fourth study began by tinkering with images of various local cityscapes (adding vegetation to places where it is currently lacking) before seeing how Chileans responded to these pictures (Navarrete-Hernandez and Laffan 2019). The authors took such an approach based partly on the argument that, even though a great deal of work has focused on the visual experience of parks, many cities cannot boast these facilities. Their argument is consistent with the findings of Hartig et al. (2014), who note how parks have been the predominant focus when researchers have thought about what they should do with the suggestion that greenspaces promote public health.

But what if, for other reasons altogether, and which have comparatively little to do with effective greenspace provision and design, people are becoming disinclined to derive these benefits? What, for example, about broader processes of cultural change: the trends that gradually push us to live our lives in some ways instead of others and which, often without us necessarily noticing, are quietly shaping the future of greenspace experience? Scholars occasionally argue for the need to consider such broader sweeps of change. Grinde and Grindal Patil (2009), for example, pursue the contention that, though greenspace benefits appear to exist, we must still stay mindful of their 'penetrance'. Their point is that we should not forget how various cultural factors may very well be over-riding their apparent draw. Hartig (1993) has similarly argued for studying greenspace experiences in a 'transactional perspective', namely alongside, rather than apart from, the broader processes that either push people towards or away from these experiences. His idea is that, though positive responses may be hardwired into humans, the likelihood of different groups seeking out the experiences that produce them is another matter. If spaces containing certain kinds of living vegetation are where we feel most at home, we might imagine that tempting people into such environments shouldn't be so hard. Not so, according to some others.

The Extinction of Experience

Enter the 'extinction of experience' thesis. This is the idea that, despite the various apparent benefits of spending time in greenspace, many lives are increasingly decoupled from regular outdoor experiences with living vegetation, different forms of local animal life, and other natural features. According to Soga and Gaston (2016), fewer and fewer of those who live in modern societies are having enough contact with the natural world. This, according to Pyle (1993), the originator of the term 'extinction of experience', sets up a vicious circle of increased alienation from experiences that may very well be beneficial to us, but to which we could be increasingly indifferent – a cycle of growing disaffection that may well have, according to many of these researchers, some fairly disastrous consequences. Zooming out to contemplate the broader history of humankind, Kellert (2002, p. 118) goes as far as to argue that modern US society has 'become so estranged' from its natural origins, that it now fails to recognise its 'basic dependence on nature as a condition of growth and development'. It's easy to see the problem here. If many people no longer care about, or see themselves as part of, the wider 'natural world', humanity could very well be drifting towards a rude awakening, whilst (adding insult to injury) being comparatively unhappy along the way by virtue of how people are increasingly oblivious to the benefits that flow from greenspace experience.

This is an alarming prospect. And we should examine the processes involved before we abandon all hope. The leading villain in this story is often urbanisation. Despite the best efforts of some of the park planners and researchers discussed above, city living is often taken to draw people away from the likelihood of beneficial encounters with greenspace. If the vast majority of humans are now living urban lives, researchers should examine how everyday experience is structured in different cities around the world and see what that tells us about the likelihood of people venturing out into greenspaces (see, for example, Turner, Nakamura, and Dinetti 2004; Fuller and Gaston 2009). Another anxiety centres on how new recreational activities could be replacing outdoor play. The migration of social life online and the ways in which many children are coming to prefer computer games over outdoor activities has been a particular source of worry for some (Pergams and Zaradic 2006; Soga and Gaston 2016). Just how busy many people now are occasionally gets a mention – how it is that many groups, in cities at least, now feel themselves to be too rushed to think about ways of inserting more greenspace experience into their lives (Lin et al. 2014). Ward Thompson (2002) develops this last point by considering the apparent stigma of lingering without purpose within societies whose members feel they should be seen to be doing something. Could it really be that the simple idea of sitting and contemplating greenspace has become too challenging for those who feel they ought to be otherwise preoccupied? This connects to concerns (Duvall and Sullivan 2016) about how our technologies can stop us from reaching the point when we are able to derive greenspace benefits even when we have managed to get there. Smartphones might provide a helpful social crutch if we find it difficult to appear purposeless in a park. But, if we have made it to the park but cannot help but look at our screens when we are there, is being there really doing us so much good?

Others have pointed to how an alternative series of, less frequently discussed but no less important, social trends have also served to discourage people from acting on the suggested desire to be around plants and trees. Bixler and Floyd

(1997), for example, make the obvious but crucial point that, if we stopped for a second and allowed ourselves to consider changes in how human lives are most commonly organised (instead of jumping the gun with a premature focus on effective landscape design), we should be unsurprised to see a growing separation between everyday life and outdoor greenspace. Because of how societies have set about making life easier for themselves, natural areas may now be 'uncomfortable' for many. As they noted, in the twentieth century, most advances in home design have sought to improve comfort (see also Shove 2003, on these trends). So, whilst central heating and air-conditioning, showers, sinks and other inventions may initially seem like fairly innocuous and attractive technological advances, they have probably, according to them, also resulted in a 'narrowing of comfort range and lowered tolerance for a wide range of environmental irritants' (p. 448). In developing this suggestion, theirs is a very different way of seeing human encounters with the 'natural world' when some of the above studies can tend to celebrate greenspace benefits. Could it actually be that many people now see outdoor greenspaces as places of environmental 'irritation' (more than enjoyable restoration) when compared to their indoor comforts? Perhaps we should consider what keeps people away from outdoor greenspace as much as what they would ideally experience if they went.

On that point, others have emphasised the importance of acknowledging the continued geographical bias in studies of greenspace benefits. This has led researchers to overlook certain important parts of the puzzle. Specifically, because many studies have been done in relatively temperate climates, the outdoor discomforts that are likely to be more keenly felt elsewhere in the world are often downplayed (Keniger et al. 2013). In other words, these studies tend to picture 'the outdoors' as a pleasant environment in which to linger such that those who do so will soon start to reap the restorative benefits provided by greenspace. Sometimes this is even part of the research design when studies have attempted to control for these 'contextual' climatic matters in order to study the effects of spending time with greenery in a more scientific way (see, for example, Bamberg, Hitchings, and Latham 2018). Yet, in very many cities around the world, it is often simply too hot, too cold, too sticky or too windy to make it an attractive proposition to sit outside and start accruing the benefits that feasibly flow from living vegetation in parks. It is a straightforward, but no less important, point that, if the people involved are rained on, or they start to sweat, they might soon leave (and potentially resolve never to spend time in such 'irritating' environments again).

If we turn to a different reading of the persistent effects of our evolutionary past, we are encouraged to see another set of reasons to be reticent about lingering for too long in these spaces. Whilst humans may very well be fundamentally attracted to particular vistas and the presence of greenery, we should not forget how there have often been challenges and threats concealed within.

These range from spiders and snakes to irritating plants and stinging insects (Bixler and Floyd 1997). It might therefore make good sense to recoil from these environments and retreat into the sanctuary that was once provided by caves and other forms of basic shelter and is now more commonly found in houses and apartments. Others have developed this thinking by turning to how greenspaces can feel like unpalatable places of 'risk' such that many do not go to them because of a background sense that they are insufficiently safe or, returning to the less intense feelings of aversion that Bixler and Floyd point towards, insufficiently sanitised (Skår 2010). Another study has considered how we might feel more relaxed if we can see for a good distance without potential assailants seeing us – viewed in this way, being immersed in vegetation that can also conceal threats is understandably unappealing (Gatersleben and Andrews 2013).

How to Respond?

What should be done about this? If we accept that there is more to this issue that providing attractive urban parks, what other solutions are there? One novel response is to think about how comparable experiences and benefits could be provided indoors. Could getting people to look at greenery on screen (or experience it through virtual reality) have the same effects? Perhaps for older people in ageing societies this could be a particularly good idea when the real-world equivalents can be physically daunting for this group (Depledge, Stone, and Bird 2011)? However, such a strategy could also push those involved even further away from the outdoors by giving them everything they need from nature inside. Presenting dramatic natural environments on screen might furthermore make the local outdoor reality increasingly dull by comparison (Ballouard, Brischoux, and Bonnet 2011). Building on the idea that we need to engage with, rather than ignore, the changing ways in which people are living in cities, another suggestion is that policymakers might do better to focus on making greenspace easier for people to encounter without making active choice to go to parks and gardens. Perhaps we should focus on the 'incidental' interactions associated with where they already walk, work and live (Cox et al. 2017a).

In a study that suggests those Australians who go to urban parks are doing so because of their personal affinity with these places more than the proximity to their homes – what they call the tension between 'orientation' and 'opportunity' – the logical conclusion is that we should encourage the affinity (Lin et al. 2014). For these authors, that means potentially undertaking a kind of 'nature awareness training' for young people so that this affinity is established in these early years. The hope is that this will stand them, and wider society, in good stead as they grow up. Indeed, children have been a particular target for this kind of argument, connecting to anxieties about what others have called 'nature deficit disorder' (Louv 2005) – the idea that, because many modern children don't play outdoors as previous generations apparently did, they are already suffering as a result. Soga and Gaston (2016), for example, float the suggestion that parents should perhaps be making the effort to force their children outside (once there, they'll soon get used to it, and soon start to like it). Could that eventually turn the tide on the broader cultural turn away from greenspaces that these studies worry about? And if we succeeded, as a number have considered, then benefits may not only be accrued by the individuals involved. Indeed, there is, in fact, some evidence that the result could be a greater sense of care for the natural environment, a stronger commitment to conservation and an increased interest in the health of the planet. Staying with the focus on contemporary young people, if one of the biggest challenges relates to how attractive 'screen time' has become to them (Larson et al. 2018), perhaps smartphone apps could help (Dorwood et al. 2017)? Either way, the concern here is that, if many young people are increasingly cocooned from outdoor experiences, they could quite easily become unaware of what is happening in the wider environment at a range of scales (from global climate change to local biodiversity loss). And soon that could be too late to fix.

It can be tempting to see young people as the obvious focus for attempts to tackle this problem (in the hope that they will somehow escape the challenges currently faced by the rest of us when they grow up). Indeed, the whole discipline of environmental education is essentially predicated on this idea. Within it, and regardless of where wider lifestyles seem to be headed in many places, it has become quite common to buy into the suggestion of 'getting them early' and then hoping for the best (Collins and Hitchings 2012). Yet, it is entirely possible that today's young people will be socialised into future societies that are even less inclined to linger in greenspaces, irrespective of our attempts to get them bitten by the greenspace bug in their relative infancy (Asah, Bengston, and Westphal 2012). Indeed, if we think life course is important, perhaps we should consider how people move through other stages that each present their own opportunities and challenges in terms of establishing an affinity with the natural world (Bell et al. 2014). Then there is the much-vexed matter of how some ethnic groups feel that public greenspaces are not really for them, partly because they often congregate in parts of the city where they are comparatively uncommon (Gentin 2011). We have also seen studies exploring how women have particular ideas about the forms of urban greenspace in which they feel sufficiently safe and comfortable (see Krenichyn 2004). Others have also considered how those living in disadvantaged areas may particularly benefit from nearby greenspace (potentially acting as a kind of buffer to dissipate the stresses of experienced financial hardship) (Ward Thompson et al. 2016), and how older people might feel that they gain as much from viewing greenery from their homes as going out into it (Day 2008).

One recently popular way of thinking about encouraging greenspace benefits has been to speak in medical terms and to talk of the most effective 'dose' of nature experience to foster individual and collective health (Gladwell et al. 2013; Cox et al. 2017b). This is not without its problems in terms of downplaying variable circumstances (how groups might respond differently to their dose and face different dosing challenges) (Bell et al. 2019). Yet, for me, this is an apposite way of thinking about the issue because, when we are taking our medicine, we are doing something that we know is good for us, but which we can otherwise easily overlook. This is the essential idea that justifies the focus of this book. Within it, my aim is to consider how certain outdoor experiences that may feasibly involve beneficial encounters with plants and trees might be squeezed out of everyday life. My thinking is that we can make urban greenspaces as attractive as we like. And (without being too dramatic about it) we can extol the restorative benefits that come from spending time in these spaces until we are blue in the face. But, if many city people are being captured by certain patterns of everyday living that render them oblivious (or, perhaps more rightly, incapable of responding) to the benefits of being with trees and plants, the mounting evidence suggesting that going there could do them much good will be of little effect.

With that suggestion in mind, this book turns to a variety of situations that may initially seem trivial (I'll make no bones about it). It will spend time attending to how a sample of city lawyers speak about 'stepping away from their desks' and how some recreational runners have ended up on treadmills. It will explore why the basic idea of living plants can prove challenging for some of those who are lucky enough to own a domestic garden and how young people feel they should wash at summer music festivals. The processes at play in these situations are those to which even the people involved may give little thought. Nevertheless, my argument is that they could eventually end up having significant consequences. But I am getting ahead of myself here. The next step is to discuss how I became interested in this topic and the concepts on which this book draws to explore it.

The Nuts and Bolts of Nature

I'm a geographer. And the reason why I became interested in this topic is partly because, in recent years, some of those working in my field have been pioneering some original ways of looking at human experience that I figured could be helpful here. My thinking was that, if we stand to benefit from a closer examination of greenspace experience in everyday life, they had something useful to say. This is because a number of my colleagues have become increasingly focused on the detail of how people and environments interact. This fits with a longstanding focus (some would say this is what defines a geographical approach) on how human societies and physical systems come together in specific contexts (and how these relationships change over time) – the kind of processes that can often end up lost in the cracks between disciplines, which have been more avowedly focused on either 'social' life or the 'natural' world. In recent times, this branch of geography has become especially interested in how exactly that 'coming together' happens within particular encounters in particular places. This has been an exciting time to be a geographer working on 'nature–society relations' as an expanding menagerie of creatures and concepts has been called forth in our conferences and articles in an attempt to get to grips with how exactly these relationships take shape (see, for overviews, Ginn and Demeritt 2009; Castree 2014).

The approach from human geography to which this book is indebted stems from how some of my colleagues have sought to think afresh about how the nominally 'natural world' is best studied. I've put it in inverted commas now because many of these scholars have been increasingly suspicious of the term. This is partly because 'nature' is such a powerful concept (think about how when we say something is 'natural' it suddenly becomes quite hard to argue against) in a way that makes it worth questioning how that power is wielded in different contexts. It is also because as soon as something is labelled as part of 'nature' it immediately becomes imbued with certain positive qualities that might not always apply. Few would say that they don't like 'nature' because of these associations. However, even though we may like to think that we appreciate 'nature' (and linking back to the different ways of characterising greenspace experience highlighted above), when out walking in the woods, for example, were we to be suddenly stung by a bee, we might find ourselves appreciating it rather less. With such examples in mind, the contention of some of my colleagues has been that it is not at all clear that the various phenomena we often find ourselves lumping together as 'nature' have all that much in common at all. Perhaps we might do better to sidestep the idea of 'nature' altogether and instead look afresh at the various phenomena that were previously subsumed under this unhelpfully general heading. Doing so, many have now argued, allows us to get a better handle on how exactly people live with the different 'entities' involved (or the 'nuts and bolts of nature', if you like).[3]

There has been a keen interest in animals here. This is partly because this work has focused on exploring the individual capacities of creatures in ways that were previously downplayed when they were unhelpfully bundled together and seen as belonging to 'the natural world' – namely their ability to act, to make their presence felt, to do things that we might not always expect or want. If this was the suggestion that these geographers wanted to acknowledge and explore, animals presented an obvious focus for their studies because their 'agency' was immediately apparent. In other words, animals are clearly and self-evidently 'alive' as individual lifeforms. And they have accordingly served geographers well in exploring the truth of these claims: how people manage an octopus in an aquarium in North East England

and how the octopus itself has a hand in fostering certain relations (Bear 2011); how the actions of certain birds help us to understand the practised appeal of birdwatching (Lorimer 2008); and the specific cultural narratives associated with sharks and how well that matches up to the reality of co-existence in Australian waters (Gibbs and Warren 2015). These are just a few examples from the growing subfield of 'animal geographies' (see Gibbs 2019, for a recent review).

If we were to start questioning 'greenspace' in this way, the first thing that we might do is to set about smashing this rather broad idea into pieces so that we can start our inspection of its components in earnest (or, as Phillips and Atchison (2018) nicely put it, we should make the effort to 'see the trees' for the forest). In other words, what some of those working in this field would immediately ask is what is this 'greenspace' idea composed of in terms of its physical materials and how exactly do people handle specific elements? By thinking in the comparatively distanced, and predominantly visual, way implied by the very idea of 'greenspace', these geographers would worry about how we may be missing out on much of how it actually is to experience greenspaces. Perhaps we should examine trees as physical, growing, living individuals – as dynamic creatures that provide shelter, fruit, leaves, opportunities to climb, hide, and to gather people around them (Jones and Cloke 2002). In this sense, they are like the above greenspace researchers in that they are interested in how people respond. The difference is that they would explore these issues by looking at how exactly life goes on in specific contexts. Another strategy would be to allow our attention to drift down to the ground and consider the ways in which people live with plants. This has been the subject of some geographical interest, sustained in part by colleagues who have set out to emphasise how plants have distinct capacities (that are different from their more evidently active animal cousins, but nonetheless there). They point to what they have called the 'vegetal politics' (Head et al. 2017) of how we manage plants in contexts that range from vine growing to weed control. This book draws inspiration from this work in terms of looking closely at lived experience with components of the nominally 'natural' world.

Entangled and Disentangled

But there are also ways in which it takes a different path. As mentioned, one of the defining features of this work has been a commitment to looking at how 'social' life is never entirely social. In other words, part of the point has been to recognise how people must contend with all sorts of materials and forces in their lives, even though a great deal of previous research tended to downplay these features (with the 'social' sciences looking at people and the 'natural' scientists looking at physical processes). These geographers have been keen to demonstrate how humans are not so separate and apart from the components of the natural world as we (rather arrogantly) might have been inclined to see them. And so, to use two early landmark examples from this field of work (Whatmore 2006; Hinchliffe 2007), their aim was partly to provide a new perspective on how human life goes on. But it was also to determinedly see it differently – ultimately to provide accounts in which people are shown to deal with a variety of materials, animals and plants in ways that they may not always want. So this work was also wrapped up in an ethical project of, in effect, bringing us down to earth (Whatmore 2006) by being a little more humble about the importance and power of our species. A similar objective was to 'animate' the material world (Hinchliffe 2007) by belatedly seeing it as a more central character in the story of how social life goes on.

The key point is that this work sets itself the dual task of both recognising that nature's components can act into the social world, but also, and crucially for me, encouraging us to look at things in this way. For example, one of the ways in which those working in this field have increasingly imagined how human life goes on is in terms of 'entanglement' (Harrison, Pile, and Thrift 2004; Jones 2009). This has become a popular term partly because the 'anthropocene' demands that we see ourselves as entangled (Hamilton 2017) since the idea of an external nature no longer makes much sense if we have entered a new geological epoch defined by human 'impacts' on the earth. Some recent examples of geographers encouraging us to see society as 'entangled' include Robbins (2019), who considers how this idea can help us reimagine standard scientific practice, Gibson-Graham, Cameron, and Healy (2019), who use it to question common ways of seeing manufacturing, or Morris (2019), who draws on entanglement to challenge predominant conventions of animal conservation. These researchers have been drawn to this terminology because part of their intention is to emphasise how individual people are constrained in terms of what they can do with nature's components – that they are subject to the willingness of various lifeforms, environments and materials to bow to the wishes of the humans with which they live. There is also a nicely suitable organic image that is conjured up here – life is a project in which humans must respond to the reality of their existence amidst a thicket of other agencies.

The suggestion that the geographer's role is one of rooting social life more fully into the material world has also influenced the people who have been studied using these ideas. Often these have been those best placed to help us develop this approach by telling us about the benefits of acknowledging their entanglements. To give three examples of recent work in this vein, we have seen some groups of English farmers recognising the benefits of recalibrating their relationship with the soil in a way that attunes them to how they should manage it in ways that are not always so controlling (Krzywoszynska 2019). Another example asks us to attend to how 'off-gridders' in Canada can take pleasure from being required to live within the limits of what variable weather conditions provide to them as part of a broader ethical commitment to reducing their impact on the planet by consuming less energy (Vannini and Taggart 2015). Returning to greenspace, a third example relates to a study of Australian city residents who, when asked by the government to report on the health of their local parks, wrote love letters to their favourite trees (Phillips and Atchison 2018). These researchers have given us some arresting and often life-affirming accounts of how certain groups of people are responding to some difficult environmental times. But those who are studied here are also those whose personal sentiments often chime well with a broader ethical project of seeing humankind as entangled.

I have often wondered about how, in many contexts, people seem quite happy to live some relatively disentangled lives. Indeed, they might even prefer that (in view of how being entangled instinctively seems unattractive, it is perhaps surprising to see that it has become a kind of rallying call for attempts at reimagining social life). My thinking here is that, though it has been tempting to focus our studies on those who see themselves in this way, this leaves broader questions about the rest of us open. Wider societies might not want, or have the time, to become entangled. Going back to how Bixler and Floyd (1997) noted how increasingly sanitised lives could be engendering new levels of reticence when it comes to encountering the 'natural world', they were effectively alerting us to how modern societies have been quietly disentangling themselves. Kaika (2004) argues something similar when she highlights how it can now feel 'uncanny' to be reminded that constant domestic water supply, for example, ultimately depends upon what the 'natural world' is able to provide. Ingold (2004) similarly points to how hard many societies have worked to achieve standards of 'modern metropolitan' living that are all about achieving a state in which their members are relatively oblivious to these kinds of entanglement. As he points out, many people now give little thought to the practical challenge of urban walking, for example, partly because their societies have furnished them with shoes and surfaces that help them to forget about it.

This takes us back to the extinction of experience thesis. My point now is that, whilst this area of geographical work has trained our attention onto how exactly people handle the 'nuts and bolts' of nature (and whilst doing so has breathed new life into the accounts that we write about how people live with the nominally 'natural world'), less has been said here about the broader sweeps of social change that may quietly be serving to pull people away from them. This book aims to connect the two. In order to do so, it now turns to a different way of seeing human life that could help me to explore how people may be turning their backs on greenspace experience, irrespective of whether they would, in principle, like being there.

Acknowledging the Power of Practices

In order to explore how broader processes of social change might be reconciled with my interest in the detail of how human relations with 'nature' play out, I turned to another body of research. Here the focus was more squarely on patterns of 'everyday life', how those patterns are sustained, how they are experienced and how they evolve. Unlike how some of the geographers were keen on finding entanglements, this work was interested in how predictable routes are carved out for people. These were theories of 'social practice' and they have attracted a growing amount of interest in recent years. There are many versions, with some being preoccupied with change, others with the role of materials in sustaining practices, and others with the most conceptually sophisticated vision of social life. There are also several overlaps between these ways of imagining everyday life and those to which the above geographers were drawn (see Maller 2019). In view of that, I should itemise the components that I picked from the expanding menu of suggestions produced by this second body of work (Shove, Pantzar, and Watson 2012; Hui, Schatzki, and Shove 2017).

One key feature is there in the basic terminology. The idea here is that we should imagine social life as essentially an outcome of how 'social practices' come about and evolve. The implication is that we should focus on how different practices, namely particular recognisable activities, spread through societies, and how they draw people into this process as time goes on. In this respect, a central aim of this work has been to push beyond the way in which it was previously tempting to paint social life as either governed by imagined 'structures', like culture, which were ultimately unsatisfying and unhelpful when, in the final instance, they made individuals seem entirely without choice, or instead to invert the picture by championing the 'agency' of people. This was also deemed unsatisfactory in the sense that it could end up seeing them as in a constant process of making decisions about how they wanted to live instead of attending to the constraints that curtailed their actions.

Recognising that there was probably some truth to both claims, and in an attempt to reconcile these two visions of social life, 'practices' were seen as providing us with a promising path between them (Giddens 1986; Reckwitz 2002). For me, they were also promising in the way that they pointed to how people were sometimes reflecting on their actions and sometimes simply doing what the situation (or rather the practice) encouraged. People are reimagined here as potentially little more than the mere 'hosts' or 'carriers' of practices (Shove and Pantzar 2007), who, once they have been effectively infected, may have little occasion to question certain actions thereafter. This idea chimed well with the above suggestion that a range of, hitherto relatively unacknowledged, social trends could be creating new ways of relating to greenspace.

If I were to draw on these ideas, my focus should be on certain practices, namely activities in which many people commonly take part. Various practices have already been examined using these ideas, including how people travel around (Watson 2012), how they organise their eating (Warde et al. 2007), how they keep warm at home (Gram-Hanssen, 2010) and how they are drawn into particular leisure activities (Shove and Pantzar 2007). Building on that work, this book is concerned with four practices that are currently commonplace and which could be feeding into the 'extinction of experience' that some of the above researchers have worried about. Accordingly, the aim is less about positioning people as essentially 'entangled' in the sense that they are grappling (in presumably at least partly conscious ways) with how they should respond to specific material circumstances. Rather it is more about how familiar settings, along with the accumulation of experience, can serve to do almost the opposite – namely curtail the likelihood of much active reflection by providing relevant groups with conditions that structure their actions. We might, as we will see, sometimes reflect on the processes involved (how did we find ourselves in this situation?). But practices can equally discourage those involved from too much analysis.

Purpose and Restoration

For me, theories of social practice suggested a useful way of studying how easily outdoor experiences infiltrate everyday lives. And whilst there are many ways of accommodating the nuts and bolts of nature within theories of social practice, my starting point was to see these components as potentially destabilising threats to certain everyday practices that are both common and often increasingly widespread. Turning again to the 'entanglement' metaphor, this encouraged me to see outdoor greenspaces as environments that people might feasibly get caught up in in ways that might snap them out of their preoccupations.

Though their authors would be unlikely to express it as such, some of the above greenspace studies that I started with also hinted at a certain kind of entanglement. One of the suggestions that they examined, after all, was about how greenspace experience could be beneficial because it allows us to transcend our immediate concerns and put aside what was troubling us beforehand. In effect, people can become 'mentally entangled' in a way that leads to valuable forms of human respite and restoration. My point is that, whilst this now sounds like an enjoyable experience, it may also, for many people, feel like a risk if it proves hard to return to everyday life afterwards. The assumption that runs through some of this work is that it should be relatively easy to go back refreshed (after a beneficial burst of greenspace restoration) to what was happening beforehand. But the truth of the matter might be another thing and, if we run with the argument about starting with the practices, it is also possible that our practices might not always be so willing to let us escape their grip.

This finally takes me to the 'subjectivities', or personal feelings, associated with carrying out practices. Understandably, these features have not often been at the forefront of analysis for those working with these theories. The point, after all, was partly about putting the practices, rather than the people, centre stage in our understandings of how patterns of everyday life become established and evolve. If that is the aim, too full a focus on the experiences of taking part in relevant practices risks analysis drifting back towards a more 'people-centred' account when part of the point was to position them as not always so in control. One of the most popular ways of describing how it is to carry out a practice, however, is with reference to Schatzki's (1996) idea of a 'teleoaffective structure'. This idea draws our attention to how, in the course of carrying out relatively familiar activities, the practice effectively carries us along towards its usual end point (or until we have to stop). In effect, we are swept along by the practice. This is the idea that I want to take forward here. Schatzki is drawing our attention to the feeling of purposeful flow that we may experience when we carry out a practice, when we are getting on with things, getting things done, just as we usually do. There is tension here between purpose and restoration. It hints at the challenge of greenspace experiences infiltrating certain ways of meeting the demands of everyday life that might be becoming more widespread and which encourage those involved to act and think in certain ways. For many, the whole idea of 'connecting to nature' is motivated by the desire to 'disconnect' from the presumed pressures and stresses of modern urban living (Kaplan and Kaplan 1989). But the extent to which it is easy, in the moment, to achieve such 'disconnections' is, I would argue, a matter for investigation.

Establishing a Focus

The above discussion has covered what this book will attempt to do, why that could be a worthwhile undertaking, and the ideas on which it draws. Based on that, these are the basic ingredients for the approach being developed here:

1. **Environmental estrangement** From the existing research on greenspace benefits a number of ideas have been taken. First, the whole rationale for this book is derived from one of the essential findings of this field, namely that spending time in the company of trees and plants can provide people with a diversity of benefits, from mental restoration, to lowered blood pressure, to reduced stress and greater feelings of self-esteem. And, more than that, this might then even spill over into a broader sense of care for the living environment at a time when encouraging such commitments might be more important than ever. This work also suggested that many urban lives were, despite all these potential benefits, now unlikely to involve so much time outside with plants and trees. A variety of potential explanations were offered for this. Less had, however, been said about how a kind of environmental estrangement might be observed and examined in everyday life.

2. The everyday outdoors Meanwhile, in human geography, I found an eagerness to examine how exactly different groups live with the 'natural world' and an interest in seeing social life as a negotiation with physical phenomena. These ideas have informed my focus on the everyday outdoors. Here I will partly follow their lead by paying particular attention to the specifics of the situations at hand and exploring how the various 'natural' materials involved are managed differently according to those specifics. The 'outdoors' commonly figures as a self-evident, unremarkable background feature of life. My aim here is partly to turn the tables on this situation by considering how the 'outdoors' is home to a whole raft of phenomena – living plants, changeable weather, different kinds of dirt – that deserve examination. This is what greenspace experience often involves, after all, and we should examine how these phenomena are allowed to complicate human lives at a time when the assumption that people are best kept away from them is often quietly gaining ground.

3. Unsettling practices Yet, just as some of my geography colleagues were seeking out and celebrating the capacities of various creatures and forces that were previously swept under the social science carpet, I also wanted to examine how people can be recruited into the reproduction of everyday practices in ways that meant these same capacities might fade into the background of their lives. So I turned to concepts of 'social practice' because they promised an appreciation of how the 'extinction of experience' was practically achieved in ways of which the people involved might only be partly aware. The key idea that I wanted to take forward from this work was therefore about seeing society as comprising practices that can recruit people in ways that, as we will see, might effectively distance them from certain beneficial outdoor experiences. In this regard, the value of these concepts stems from how they highlight the subtleties of these processes in which relevant groups figure as sometimes able to question, and potentially amend, their actions, and sometimes effectively controlled by the practices that live through them. In that respect, I wanted to examine how, in different contexts, a range of potentially unsettling environments were handled – when and how do people submit to the practice? And when and how might they struggle free in ways that take them towards a fuller relationship with the outdoors?

The Title and the Plan

The book title is an attempt to combine all the above components into a succinct form of words to guide my endeavours in the following chapters. But I'm also quite keen on the 'unsettling outdoors' because of how it reframes the way in which societies like mine often think about how certain environments are, and should be, experienced by their members. All around the world, it is common to speak of the 'great outdoors'. Imbued with various beliefs about healthy exercise and the personal regeneration that comes from escaping the stresses of city life, it's an immediately appealing idea. It is also one which often pops up in our ongoing societal conversations about where people should go and what they should ideally be doing. And, to return briefly to some of the research findings with which this chapter started, we might agree that greenspace experiences could be 'great' in the sense that they do various good things to people, and in ways that might be more important than ever. Yet, as I have also argued, what many people find when they go outdoors may now feel somehow strange and unfamiliar to them. Indeed, it is exactly because of these attributes that it could be fairly tempting for certain groups to shun the outdoors in their everyday lives even though they may very well know that adjusting to it could be good for themselves, for others, and for the wider environment too. In short, though 'the outdoors' may be 'great' in principle, experiencing it can also be 'unsettling' in practice.

The Location and an Overview

My topic is now fixed. What remains are questions about how to go about exploring it. Building on the suggestion that the problem partly stems from the relatively recent human migration to cities, I take everyday life in one city as a focus. This is London. London is a city that contains various social practices that may be serving to separate pattens of everyday life and the possibility of beneficial greenspace experience. Though there is a relative abundance of greenspace in this city, I should also emphasise upfront that I do not want to imply that, as soon as Londoners venture outside their buildings, they will immediately step into a restorative paradise teeming with healthy trees and plants. Rather my idea is that, in order to understand the likelihood of greenspace benefits infiltrating everyday lives, both in this city and elsewhere around the world, we can benefit from stepping back from the greenery focus and turning to how outdoor environments are handled by those with an established relationship with particular practices.

Four practices are considered and four characterisations of how people ended up relating to the outdoors are offered. First, I consider the practice of office work, how that has been studied with regard to greenspace experience and how those currently involved can end up 'forgetting' about outdoor environments nearby, whether green or not. Second, I turn to the practice of recreational running, how running has been connected to greenspace benefits, and why those who currently run indoors on treadmills do so partly because they are 'avoiding' exactly those outdoor environments in which they feel running should ideally happen. Third, I consider how well the idea of shopping fits with how Britons handle their domestic gardens and how, despite the ways in which some London garden owners are unsettled by the suggestion of living plants, they can end up 'succumbing' to the pleasures of a less controlling approach. Finally, we take a short trip away from the city to consider what has been said about the benefits of camping in more 'natural' environments before turning to the practice of showering and the ways in which young people (who otherwise wash more than most) cope with the challenge of a summer music festival. Here I examine how some of them can end up 'embracing' the dirt and discomforts of a shared experience outdoors in greenspace away from town. Together, I will argue, these processes (of forgetting, avoiding, succumbing and embracing) deserve more attention, despite rarely being the focus of those aiming to foster greenspace experience.

But before that, I want to say more about research methods since this book is also about strategies for studying social life. These are my concerns in the next chapter. Within it, I make a case for attending to how people speak of the social practices they are involved in carrying out and how practices can encourage people to speak in certain ways. I will also say some more about where we will go to explore environmental estrangement in the four case study chapters that follow. But, for readers who are less interested in research methods, it is possible at this point to skip forward to the case studies. As we go through those, the book will move from spaces of work, to spaces of exercise, to spaces of leisure, both at home and away. In this respect, it gradually turns to social contexts in which we might imagine there to be increasing amounts of time and inclination to revel in enjoyable and beneficial engagements with outdoor environments. These chapters will consider the extent to which this is the case before the book ends by drawing a series of broader conclusions about how the social future of greenspace benefits is investigated and influenced.

Endnotes

1 It should be acknowledged that these studies commonly work with the assumption of all people sharing the same essential response to stimuli. In this way, they downplay the likelihood of responses varying according to the cultural and demographic characteristics of individual groups. Nonetheless, and notwithstanding how there will be variations of this type, when taken as a whole, these studies still provide an impressive weight of evidence.

2 Jorgensen (2011) also considers how the whole idea of 'landscape design' is fundamentally predicated upon the implicit prioritising of visual experience.

3 In this regard, many of my geography colleagues are naturally suspicious of some of the arguments presented earlier in this chapter. In their thinking, the problem with many of the psychological studies that attempt to explore how 'humans' experience greenspace is that they don't properly engage with context, namely with variation between groups and places and change over time. This reticence is understandable in view of the different purposes of these two disciplines (one often hoping to understand what is believed to be an essential human response; the other more interested in exploring how things work out differently in different situations). However, as I see it, it is a shame to ignore the findings

of relevant studies because they do not share the same starting assumptions. After all, whilst we may be arguing about the best ways of defining the problem, things could be changing fast in wider society (and finding the most effective responses to these changes will surely come from consulting studies undertaken under a variety of disciplinary banners). An alternative way of characterising the benefits of a geographical approach is about being willing to combine different insights in pursuit of a fuller understanding of the issue at hand. That is a much nicer fit with how I like to think about the approach being taken by this book.

References

Asah, S., Bengston, D., and Westphal, L. (2012). The influence of childhood: operational pathways to adulthood participation in nature-based activities. *Environment and Behavior* 44: 545–569.

Ballouard, J., Brischoux, F., and Bonnet, X. (2011). Children prioritize virtual exotic biodiversity over local biodiversity. *PLoS ONE*. https://hal.archives-ouvertes.fr/hal-00613629 (accessed 26 January 2021).

Bamberg, J. Hitchings, R., Latham, A. (2018). Enriching green exercise research. *Landscape and Urban Planning* 178: 270–275.

Bear, C. (2011). Being Angelica? Exploring individual animal geographies. *Area* 43: 297–304.

Bell, S., Phoenix, C., Lovell, R. et al. (2014). Green space, health and wellbeing: making space for individual agency. *Health and Place* 30: 287–292.

Bell, S., Leyshon, C., Foley, R. et al. (2019). The 'healthy dose' of nature: a cautionary tale. *Geography Compass* 13: 1–14.

Bixler, R. and Floyd, M. (1997). Nature is scary, disgusting and uncomfortable. *Environment and Behavior* 29: 443–467.

Castree, N. (2014). *Making Sense of Nature*. Abingdon: Routledge.

Collins, R. and Hitchings, R. (2012). A tale of two teens: disciplinary boundaries and geographical opportunities in youth consumption and sustainability research. *Area* 44: 193–199.

Cox, D., Shanahan, D., Hudson, H., et al. (2017a). Doses of nearby nature simultaneously associated with multiple health benefits. *International Journal of Environmental Research and Public Health* 14: 172–185.

Cox, D., Shanahan, D., Hudson, H., et al. (2017b). The rarity of direct experiences of nature in an urban population. *Landscape and Urban Planning* 160: 79–84.

Day, R. (2008). Local environments and older people's health: dimensions from a comparative qualitative study in Scotland. *Health and Place* 14: 299–312.

Depledge, M., Stone, R., and Bird, W. (2011). Can natural and virtual environments be used to promote improved human health and wellbeing? *Environmental Science and Technology* 45: 4660–4665.

Dorwood, L., Mittermeier, J., Sandbrook, C., and Spooner, F. (2017). Pokemon GO: benefits, costs, and lessons for the conservation movement. *Conservation Letters* 10: 160–165.

Duvall, J. and Sullivan, W. (2016). How to get more out of the green exercise experience. In: *Green Exercise: Linking Nature, Health and Wellbeing* (ed. J. Barton, R. Bragg, C. Woods et al.), 37–45. New York: Routledge.

Fuller, R. and Gaston, K. (2009). The scaling of green space coverage in European cities. *Biology Letters* 5 (3): 352–355.

Gatersleben, B. and Andrews, M. (2013). When walking in nature is not restorative – the role of prospect and refuge. *Health and Place* 20: 91–101.

Gentin, S. (2011). Outdoor recreation and ethnicity in Europe – a review. *Urban Forestry and Urban Greening* 10: 153–161.

Gibbs, L. (2019). Animal geographies I: hearing the cry and extending beyond. *Progress in Human Geography*. https://doi.org/10.1177/0309132519863483 (accessed 26 January 2021).

Gibbs, L. and Warren, A. (2015). Transforming shark hazard policy: learning from ocean-users and shark encounter in Western Australia. *Marine Policy* 58: 116–124.

Gibson-Graham, J.-K., Cameron, J., and Healy, S. (2019). Roepke lecture in economic geography – economic geography, manufacturing, and ethical action in the Anthropocene. *Economic Geography* 95: 1–21.

Giddens, A. (1986). *The Constitution of Society: Outline of the Theory of Structuration.* Berkeley, CA: University of California Press.

Ginn, F. and Demeritt, D. (2009). Nature. In: *Key Concepts in Geography* (ed. N. Clifford, S. Holloway, S. Rice et al.), 300–311. London: Sage.

Gladwell, V., Brown, D., and Wood, C. (2013). The great outdoors: how a green exercise environment can benefit all. *Extreme Physiology and Medicine* 2: 3.

Gram-Hanssen, K. (2010). Residential heat comfort practices: understanding users. *Building Research and Information* 38: 175–186.

Grinde, B. and Grindal Patil, G. (2009). Biophilia: does visual contact with nature impact on health and well-being? *International Journal of Environmental Research and Public Health* 6: 2332–2343.

Hamilton, S. (2017). Securing ourselves from ourselves? The paradox of 'entanglement' in the Anthropocene. *Crime, Law and Social Change* 68: 579–595.

Han, K.-T. (2009). Influence of limitedly visible leafy indoor plants on the psychology, behaviour, and health of students at a junior high school in Taiwan. *Environment and Behavior* 41: 658–692.

Harrison, S., Pile, S., and Thrift, N. (eds) (2004). *Patterned Ground: Entanglements of Nature and Culture.* London: Reaktion Books.

Hartig, T. (1993). Nature experience in transactional perspective. *Landscape and Urban Planning* 25: 17–36.

Hartig, T., Mitchell, R., de Vries, S. et al. (2014). Nature and health. *Annual Review of Public Health* 35: 207–228.

Head, L., Atchison, J., Phillips, C. et al. (2017). *Vegetal Politics: Belonging, Practices and Places.* London: Routledge.

Hinchliffe, S. (2007). *Geographies of Nature: Societies, Environments, Ecologies.* London: Sage.

Hoyle, H. Hitchmough, J., and Jorgensen, A. (2017). All about the 'wow factor'? The relationships between aesthetics, restorative effect and perceived biodiversity in designed urban planting. *Landscape and Urban Planning* 164: 109–123.

Huang, Q., Yang, M., Jane, H.-A. et al. (2020). Trees, grass, or concrete? The effects of different types of environments on stress reduction. *Landscape and Urban Planning* 193: 103654.

Hui, A., Schatzki, T., and Shove, E. (2017). *The Nexus of Practices: Connections, Constellations, and Practitioners*. London: Routledge.

Ingold, T. (2004). Culture on the ground: the world perceived through the feet. *Journal of Material Culture* 9: 315–340.

Jones, O. (2009). After nature: entangled worlds. In: *A Companion to Environmental Geography* (ed. N. Castree, D. Demeritt, D. Liverman et al.), 294–312. Oxford: Wiley-Blackwell.

Jones, O. and Cloke, P. (2002). *Tree Cultures: The Place of Trees and Trees in their Place*. Oxford: Berg.

Jorgensen, A. (2011). Beyond the view: future directions in landscape aesthetics research. *Landscape and Urban Planning* 100: 353–355.

Kaika, M. (2004). Interrogating the geographies of the familiar: domesticating nature and constructing the autonomy of the modern home. *International Journal of Urban and Regional Research* 28: 265–286.

Kaplan, R. (1993). The role of nature in the context of the workplace. *Landscape and Urban Planning* 26: 193–201.

Kaplan, R. and Kaplan, S. (1989). *The Experience of Nature: A Psychological Perspective*. New York: Cambridge University Press.

Kaplan, R., Kaplan, S., and Ryan, R. (1998). *With People in Mind: Design and Management of Everyday Nature*. Washington, DC: Island Press.

Kellert, S. (2002). Experiencing nature: affective, cognitive, and evaluative development in children. In: *Children and Nature: Psychological, Sociocultural, and Evolutionary Investigations* (ed. P. Kahn and S. Kellert), 117–151. Cambridge, MA: MIT Press.

Kellert, S. and Wilson, E.O. (1993). *The Biophilia Hypothesis*. Washington: Island Press.

Keniger, L., Gaston, K., Irvine, K. et al. (2013). What are the benefits of interacting with nature? *International Journal of Environmental Research and Public Health* 10: 913–935.

Krenichyn, K. (2004). Women and physical activity in an urban park: enrichment and support through an ethic of care. *Journal of Environmental Psychology* 24: 117–130.

Krzywoszynska, A. (2019). Caring for soil life in the Anthropocene: the role of attentiveness in more-than-human ethics. *Transactions of the Institute of British Geographers* 44: 661–675.

Kuo, F. and Sullivan, W. (2001). Environment and crime in the inner city: does vegetation reduce crime? *Environment and Behavior* 33: 343–367.

Larson, L., Szczytko, R., Bowers, E. et al. (2018). Outdoor time, screen time, and connection to nature: troubling trends among rural youth? *Environment and Behavior* 51: 966–991.

Lin, B., Fuller, R., Bush, R. et al. (2014). Opportunity or orientation? Who uses urban parks and why. *PLoS ONE* 9 (1): e87422.

Lorimer, J. (2008). Counting corncrakes: the affective science of the UK corncrake census. *Social Studies of Science* 38: 377–405.

Louv, R. (2005). *Last Child in the Woods: Saving Our Children from Nature-Deficit Disorder*. New York: Algonquin Books.

Maller, C. (2019). *Healthy Urban Environments: More-than-human Theories*. London: Routledge.

Martin, L., Pahl, S., White, M. et al. (2019). Natural environments and craving: the mediating role of negative affect. *Health and Place* 58: 102160.

Meyer-Grandbastien, A., Burel, F., Hellier, E. et al. (2020). A step towards understanding the relationship between species diversity and psychological restoration of visitors in urban green spaces using landscape heterogeneity. *Landscape and Urban Planning* 195: 103728.

Moore, E. (1981). A prison environment's effect on health care service demands. *Journal of Environmental Systems* 11: 17–34.

Morris, A. (2019). Educational landscapes and the environmental entanglement of humans and non-humans through the starling murmuration. *Geographical Journal* 185: 303–312.

Navarrete-Hernandez, P. and Laffan, K. (2019). A greener urban environment: designing green infrastructure interventions to promote citizens' subjective wellbeing. *Landscape and Urban Planning* 191: 103618.

Pergams, O. and Zaradic, P. (2006). Is love of nature in the US becoming love of electronic media? 16-year downtrend in national park visits explained by watching movies, playing video games, internet use, and oil prices. *Journal of Environmental Management* 80: 387–393.

Phillips, C. and Atchison, J. (2018). Seeing the trees for the (urban) forest: more-than-human geographies and urban greening. *Australian Geographer* 51: 155–168.

Pretty, J., Peacock, J., Sellens, M. et al. (2005). The mental and physical health outcomes of green exercise. *International Journal of Environmental Health Research* 15: 319–337.

Pyle, R. (1993). *The Thunder Tree: Lessons from an Urban Wildland*. Boston, MA: Houghton Mifflin.

Reckwitz, A. (2002). Toward a theory of social practices: a development in culturalist theorizing. *European Journal of Social Theory* 5: 243–263.

Robbins, P. (2019). Scientific opportunities in the great entanglement. *Geographical Review* 109: 252–257.

Schatzki, T. (1996). *Social Practices: A Wittgensteinian Approach to Human Activity and the Social*. Cambridge: Cambridge University Press.

Shove, E. (2003). *Comfort, Cleanliness and Convenience: The Social Organisation of Normality*. Oxford: Berg.

Shove, E. and Pantzar, M. (2007). Recruitment and reproduction: the careers of and carriers of digital photography and floorball. *Human Affairs* 17: 154–167.

Shove, E., Pantzar, M., and Watson, M. (2012). *The Dynamics of Social Practice*. London: Sage.

Skår, M. (2010). Forest dear and forest fear: dwellers' relationships to their neighbourhood forest. *Landscape and Urban Planning* 98: 110–116.

Soga, M. and Gaston, K. (2016). Extinction of experience: the loss of human–nature interactions. *Frontiers in Ecology and the Environment* 14: 94–101.

Summit, J. and Sommer, R. (1999). Further studies of preferred tree shapes. *Environment and Behavior* 31: 550–576.

Turner, W., Nakamura, T., and Dinetti, M. (2004). Global urbanization and the separation of humans from nature. *BioScience* 54: 585.

Ulrich, R. (1983). View through a window may influence recovery from surgery. *Science*: 420–421.

Ulrich, R., Simons, R., and Losito, B. (1991). Stress recovery during exposure to natural and urban environments. *Journal of Environmental Psychology* 11: 201–230.

Van den Berg, A., Hartig, T., and Staats, H. (2007). Preference for nature in urbanized societies: stress, restoration and the pursuit of sustainability. *Journal of Social Issues* 63: 79–96.

Vannini, P. and Taggart, J. (2015). Solar energy, bad weather days, and the temporalities of slower homes. *Cultural Geographies* 22: 637–657.

Ward Thompson, C. Aspinall, P. Roe, J. et al. (2016). Mitigating stress and supporting health in deprived urban communities: the importance of green space and the social environment. *International Journal of Environmental Research and Public Health* 13: 440.

Ward Thompson, C. (2002). Urban open spaces in the 21st century. *Landscape and Urban Planning* 60: 59–72.

Warde, A., Cheng, S.-L., Olsen, W. et al. (2007). Changes in the practice of eating: comparative analysis of time-use. *Acta Sociologica* 50: 363–385.

Watson, M. (2012). How theories of practice can inform transition to a decarbonised transport system. *Journal of Transport Geography* 24: 488–496.

Whatmore, S. (2006). Materialist returns: practising cultural geography in and for a more-than-human world. *Cultural Geographies* 13: 600–609.

Chapter Two
Taking an Interest in the Everyday Lives of Others

Another Wager

If we want to encourage people to live otherwise, we might start by attending to how they speak of the practices in which they are already involved. This is the second wager of this book. In exploring it, this chapter develops three suggestions about how to examine the relationship between speech patterns and social practices. My argument is that, though there are challenges involved in asking people to speak of things that they otherwise do as a matter of course, scrutinising how they respond to such requests offers a valuable window into how patterns of everyday life are socially ingrained and potentially redirected.

The Argument

This second chapter tells the story of the how I went about exploring the processes that were introduced in the first. My ambition then was to make the case for an examination of how people can be pulled away from beneficial experiences with green environments – how, in different domains of everyday life, certain widespread practices may be serving to drive some increasingly sturdy wedges between us and them. My ambition now is to contribute to discussions about how

The Unsettling Outdoors: Environmental Estrangement in Everyday Life, First Edition. Russell Hitchings.
© 2021 Royal Geographical Society (with the Institute of British Geographers). Published 2021 by John Wiley & Sons Ltd.

everyday life is examined, and potentially influenced, by social researchers like myself. The central question that I explore in this chapter is about why I settled on the approach that I took in the four case studies discussed in this book. Boiled down to the most basic of terms, what I did was to speak with different groups of people about how certain parts of their lives went on and what that meant for how they did, and potentially might, respond to relevant outdoor environments. In one sense I was merely taking an interest in their everyday lives and, on one level, that is an unremarkable strategy. We might even imagine that people might enjoy talking in this way and be quite flattered to meet someone with such an interest in them. But, on another level, it goes against what some scholars have said about how we should put the 'social practice' concepts, with which I ended Chapter 1, effectively 'into practice'. With that in mind, the chapter starts with why there has sometimes been both suspicion and hostility about focusing too fully on the spoken word when exploring these topics. Then I offer a more hopeful account of how talking with people might feasibly work for us in this field before ending with an introduction to the four case studies.

For the purposes of this chapter, my own interest in the everyday outdoors could be seen as an arbitrary choice. In other words, though this was the topic that I was motivated to explore, the points that I make here could also inform projects that take a similar approach to others. Another aim of this chapter is therefore to identify some suggestions that could be of use both to me in my case studies and to others thinking about how everyday life is studied by social researchers like myself. They may also be of more general interest to those who wonder about how it is that we come to end up acting as we do. In terms of the material that I have collected to question these matters, this book draws on a great deal of talk: around two hundred hours of spoken exchange with various groups of people who were, at the time, reproducing the social practices that were of interest to me in my studies. We talked of challenges that ranged from thinking about how best to handle days when the air-conditioning system broke down at work, to deciding what should be done first after waking up in a festival field, to picking out the right things to put in your newly acquired garden, or considering whether, and crucially also where, your lunchtime run should take place. How people handle dilemmas such as these may strike you as rather banal. And indeed it is. But hopefully you will see by the end that this is precisely the point. After all, the power of the processes on which this book focuses stems from how certain actions can become unremarkably uninteresting, both to those involved and to the rest of us, since, once they have passed this point, they may be a whole lot harder to shake off.

In summary, my core question now is whether social researchers can learn useful things by speaking with people about the everyday practices that they are either already caught up in carrying out or some way towards a point when they are doing so unthinkingly. Without entirely giving the game away, my answer is that this question is most convincingly answered by trying, namely by attending to how relevant groups respond to such an exercise. In this way, my discussion sets the scene for the four following chapters that do exactly that.

Boxing Lessons

For many of those who either see social life as an arena for competing practices or focus on the ways in which particular practices come to capture people, the logical extension of the argument for the approach is that we should aim to observe these practices in action. We should look especially closely at what people are doing and analyse why. Potentially, we should take part in the practice ourselves, but the objective should be to get as close as possible to how it is practically done. After all, if we are most interested in how people are drawn into particular ways of routinely acting, surely the obvious thing to do would be to watch how that happens. Especially if we are sometimes encouraged to see people are mere pawns in the process (simple creatures who are passively swept up in the rise and fall of practices), it might initially strike us as a misguided undertaking to speak with them about it. They'll probably attempt to tell us 'why' they do certain things. But, in so doing, they'll likely feel compelled to position themselves as in charge of their actions (because people often like to see themselves as such) when we've already decided that it doesn't really work like that. We might go so far as to think that it's cruel to ask them to explain themselves because they won't be able to provide a full account. And that is a fair argument if the whole point of thinking about social life in this way is to recognise that people are often carried through life in a way that can be enjoyably automatic. If that is the case, they'll surely be embarrassed about publicly discovering their powerlessness. This, at least, is the kind of argument that runs through a good amount of work. Certainly, there has been some hesitancy about the idea of testing out these ideas through talk. In line with this thinking, 'observation' has become, as Halkier nicely puts it (2017), something of a 'gold standard' method for a field whose proper aim should be to examine how things are done.

   This argument is often traced back to a study in which some fairly strong statements are made about how foolhardy it is to focus too fully on talk in this field. This is the examination of his time at an inner-city Chicago gym provided by the anthropologist Wacquant (2004). In this long-term project, he set out to understand the process of being drawn into the world of boxing by becoming a boxer himself. And, drawing on this exercise, he makes some quite powerful arguments about how the successful mastery of particular modes of physical comportment are often all about the situational disappearance of talk. One of his central contentions is, for example, that the best boxers are those who cannot really tell you a great deal about what they are doing when they box. Why? Because they have trained themselves to reach a point when they are no longer really thinking about these matters. And so it is at the stage when their actions are no longer available to them through thought (and therefore no longer easily accessible to them through speech) that the boxer is ready for the fight. By then, boxing has become 'second nature', and they'll be much more likely to win because, for at least some of the fight, they'll effectively be on autopilot.

In other words, it is precisely because these sportspeople have become 'silent' (Bourdieu 1990) on certain aspects of what they do that they are more likely to succeed. Viewed in this way, the whole aim is to let the practice subsume you and not to question it since, were you to do so, all the hard work that you've put into making a series of physical responses relatively automatic would be undone. So, whilst we may be nervous of the implications of seeing practices as powerful agents that maraud through our societies 'capturing' people who are thereafter mere 'hosts' as they go, now we find some potential hosts working hard to achieve such a state. As one of the trainers told Wacquant in no uncertain terms whilst he slowly honed the craft himself, it's not good to talk too much about being 'mentally ready' for a fight. Why? Because to do so would be to overplay the mental aspects of 'readiness', when that, for these boxers, is at least as much a bodily matter. As the trainer, DeeDee, emphasises, if you are ready, you are ready, 'all the rest is bullshit, it's for the birds' (2004, p. 96). It is the alignment of mental and physical states that allows the experienced boxer to continue after being knocked out, for example. In such situations, the body effectively continues boxing on its own until the person regains consciousness. And this reticence about spoken analysis has been taken to apply to practices more generally since this study has become one of the classic statements on why we should eschew talk.

Yet what has always struck me about his study is how much Wacquant does draw on talk, usually between himself and others in the gym. In the above exchange, when the trainer is disdainful about the idea of being 'mentally ready' (it's 'for the birds'), Wacquant is clearly learning from what he says. Yes, he is doing so in a particular way (and in a particular social situation). But the understanding that is produced is still an understanding that is, at least partially, derived from verbal exchange. Viewed in this way, it seems perverse to use this study to dismiss the broader value of studying the spoken word. Clearly, it was only by asking the trainer to speak of it that Wacquant was able to explore his suspicions about how his boxers aligned bodily skill and mental state in pursuit of a prize-winning performance. Whilst, to be fair to him, he would likely have had a difficult time of it, were he to stroll into an inner-city boxing hall and straightforwardly asked them to speak openly and fully about the extent to which they were striving for an 'unconscious boxing disposition' (so there are issues here with how we organise the most insightful exchanges), taking this study to mean that using talk to examine everyday practice is misguided is surely over-stating the case. Indeed, on sifting through his written account, we see all sorts of insights emerging from the close examination of talk. We, for example, see boxers sharing gossip and 'street tales' that collectively comprise the 'hidden curriculum' of boxing apprenticeship. We learn why some recruits are deliberately branded as 'naturals'. Doing so is an interesting exercise because it renders the active analysis of the technique possessed by 'the natural' redundant (they are already doing it so well that scrutiny is to be discouraged as potentially unsettling). Either way, Wacquant is clearly learning about the process from what his boxers are saying. In other words, by looking at how and when the idea of the 'natural boxer' is drawn upon in spoken exchange at the gym, we learn about when the boxing community is willing to subject aspects of their boxing practice to a conscious form of collective analysis.

At this point you may be asking why I am talking about boxing in a book about outdoor experience. The reason is because I want to consider how I should use the concepts that I picked in Chapter 1 to study my topic. And here it emerges that one of the studies that often serves as a justificatory anchor for some of the arguments against using talk to study social practices is actually, if we look closely at what is happening within it, more a mine of suggestions about how we might embark upon and analyse the talk that takes place between us and those whose lives we thereby hope to understand. Like the boxer he aims to become, if we look at his study in this way, we immediately see Wacquant circling around particular topics in his various spoken interactions with his boxers, sparring with them and striking at the right moment with a killer question that has been neatly extracted from an appreciation of gym life that came gradually through the many hours that he spent training there himself. His is an argument for thinking carefully about how we handle talk when we study social practices, not a rejection of the idea at the outset.

Applying or Exploring Theory

It is worth saying something at this point about ways of handling social theory in research. I think this because there is a definite whiff of arrogance to the idea of dismissing talk when researching how everyday life goes on for different social groups. To reject talking with people is implicitly to suggest that it is only the scholarly researcher who can discern the true nature of everyday practices (whilst those who they study are either too busy getting on with life or simply unable to evaluate their actions as well as we are). I worry that, if I went too fully down this route, I would be seeing myself as someone who has come to assume a kind of enlightened analytical state that is characterised by a knowing and critical approach to everyday life. But I also know that I spend a good deal of my own time going with the flow of the situation at hand. And that is regardless of how I like to think of myself as a fairly questioning person when I am called upon to consider the social contexts that I am studying at the time. So, to me, this way of thinking sets up an unhelpful dichotomy that positions academics as different to others who cannot really be expected to speak of how they have become 'stuck' with their practices. Whilst it is one of the privileges of a being a social researcher to have the time to mull over the detail of processes that others may not have the same inclination to examine, it is quite a leap to suggest that they couldn't.

Though the whole point of focusing on 'social practices' for me was partly to explore how there are times when people are content, or sometimes merely resigned, about doing what the situation suggests without giving it too much thought, we should not assume that they always are because these are the ideas that appeal to us in our studies. And neither should we set off with the intention of only looking for examples that demonstrate that we can see social life in this way. To do so would be to drastically limit our engagement with people. As discussed in Chapter 1, this book was inspired by a particular set of concepts that promised a comparatively sensitive appreciation of how certain everyday practices might be pulling people away from positive engagements with the outdoors without them necessarily noticing. But how am I 'drawing' on these concepts? Or should I rather say that I am 'applying' them? What is the right verb for my relationship with concepts in this book (since the decisions that we make here have significant implications)? I ask this because, for some, the whole excitement of seeing social life in this way is that it produces accounts that are 'subversive' (Nicolini 2012, p. 23) in that they challenge the assumed centrality of human decision making and the role of active 'choice' in everyday life. If the assumption is that, were we to ask people to talk of the everyday actions through which social practices are reproduced, they will want to position themselves as having decided to act in certain ways, speaking with them about these actions is a problem because that will leave me with accounts that don't really fit with the aim of foregrounding how practices take control of us.

This puts me in a difficult position since, irrespective of what happened in the boxing gym, it rather appears that talking with people might now be undermining the whole project of seeing social life differently (by adopting a method that cannot help but reproduce the vision it was partly set up to challenge). But to think in this way is to buy into a particular notion of how researchers, like myself, should work with their theories. As I discuss next, we don't necessarily need to see them as tools that we must faithfully apply, like devotees who demonstrate their commitment by never straying from the path that their chosen concepts have magically illuminated for them. A different way of thinking about how existing theories can help us when we embark upon studies of social life is to see them as a kind of practical 'trick' – as ideas that we have up our sleeves as we set off to undertake our studies (let's see what they do for us in the field and let's learn from how well they fit with what we find there) (Becker 1998). Hui, Schatzki, and Shove (2017, p. 1) argue that seeing social life as comprising a series of social practices is a supposition that has been gaining ground in research. The key point for me here is that this is a 'supposition' – an idea that is waiting to be tested out in our studies, and one that we can potentially explore by talking together with relevant groups of people.

In the remaining chapters I will consider how human 'actorship' (Pichelstorfer 2017) features in my talk about everyday practices (how, according to the material that I've collected, people sometimes appear to be in control of their actions and sometimes not, along with the extent to which, and why, they were willing and able to evaluate these matters with me). As I see it, arguments such as Wacquant's, along with an academic interest in the means by which people are swept up in the rise and fall of social practices provide a spur for a fuller examination of how we use the spoken word to understand these processes. This book is partly my contribution to that enterprise. Within it, whilst my aim is not to use my case studies to prove my point, I should nonetheless admit that there are certain ideas that I want to explore within them. One of them relates to the links between what is 'sayable', what is 'thinkable' and what is 'doable', and how a focus on the connections between these three can help us see how everyday outdoor relations are established and sustained. A second was that the ways in which they are linked will likely depend on the social and material context at hand. Returning once more to Wacquant, it has always seemed rather strange to me to generalise about everyday life from a case in which the whole point and pleasure of boxing was about effectively losing yourself in the practice. It is unlikely to be the same for other activities and we should be wary of building broader arguments from specific cases. How people speak about sometimes unthinking actions in everyday life may be very different depending on the group, the context in which they are found, and the ways in which we talk with them.

Three Suggestions

What the above discussion hopefully makes clear is that I want to travel light with my chosen theory in this book so as to avoid being so weighed down by it that I cannot more fully engage with the specifics of the social lives that I've studied in my interviews. Having said that, I do have three suggestions that I want to evaluate in terms of what each of my four case studies has to say about them. But again, these are suggestions (some of the 'tricks' that I have to hand to help orientate my investigations): ideas to explore together with people, not a set of beliefs to impose onto the lives of unknown others.

The three suggestions:

1. Social practices create speech patterns

As a way of getting beyond the idea that people cannot speak of what they routinely do, one suggestion is that we might instead be consistent with the idea of putting practices centre stage by looking at how doing so encourages us to see the function of talk differently. According to this thinking, we should examine how practices are partly sustained by, and even become stronger as a result of, attracting certain speech patterns to themselves. Shove, Pantzar, and Watson (2012), for example, talk of how this works in sports as, partly through a process of developing a language for discussing the rules and markers of success, an activity becomes more recognisable and fixed. The idea is that, by speaking in certain ways about the practice, we act to solidify a shared sense of what the practice properly is, and that can then, in turn, partly serve to control our actions thereafter. Bissell (2014) develops a similar suggestion in seeing speech as produced by situations, rather than the outcome of an active decision to say something on the part of individuals. In his study of routine travel, he wonders whether the expressed anxieties about the stresses of a crowded commuter train are required to diffuse the tension of the situation. So, it's the commuting practice that is really doing the talking, not the person who said the words aloud in the carriage. Another comparable example is the work of Crossley (2006) on gym going. He suggests that, for too long, we have looked at this practice in terms of how people talk about what it does to their bodies, to their appearance and their health. That is part of it, he argues, but, if we really want to understand how people keep going, we should attend to what is said in passing as regulars encounter one another since such seemingly unimportant social exchanges can do important work in holding their practice in place. Like Wacquant's, his gym isn't actually all that silent and the brief interactions and nods of recognition that he notices there can, he suggests, have important effects. Schatzki (2017) similarly sees practices as 'filling out the social context', which is presumably partly done by encouraging particular forms of speech on the part of those involved. Finally, we should be unsurprised to find that Wacquant also suggests something similar in, for example, noting how his boxers talk about their training as their 'work' (2004, p. 66). Speaking in this way helps them to keep going since this phrasing effectively signals (both to the boxer and others) that the boxer (like the worker) has no choice and so, as a result of this talk, is less likely to consider skipping a session.

2. How people handle examination is instructive

Part of the reason why talking about everyday practices seems so tricky (according to some of those who have thought about this issue) is because of the way that some writers have imagined what will likely follow our attempts. For Giddens (1986), practices belong to the domain of 'practical consciousness' in which the human body simply knows what to do in given situations. Because of this helpful feature, we can get on with much of our lives in the comforting position of needing to give little thought to what we are physically doing. Then, if we are asked to question it, the idea is that the practice is dragged up into the realm of spoken deliberation. The impression is of shaking someone awake when they were hitherto happily sleepwalking though life. And that doesn't sound like a pleasant experience. This suggestion is bolstered by the linked idea that this most often happens in 'critical situations' (Giddens 1986, p. 41), namely moments when otherwise routine actions are abruptly brought into the realm of conscious thought by realising how they are out of step with the wider social scene. The implication is that reflecting on how we usually act makes us doubt ourselves (we feel silly, that perhaps we've been doing things in the wrong way all along!). And that also sounds far from enjoyable. However, though this is consistent with the point about recognising how people are not always the authors of their actions, it remains to be seen whether they would find it so hard to admit that. We may see society as increasingly impressed by the ability of its members to reflect on their actions, but whether people always share that view is another matter (Sweetman 2003). It is also the case that there may be many ways of getting people to convey a sense of how they are carried through life. Spoken exchange need not always entail such active evaluation. It could also convey a sense of how people respond to particular situations without feeling the need to question how and why they act as they do. And, even if they have little to say about certain practices because 'practical consciousness' (Giddens 1986) is the dominant force in their execution, those we hope to understand might be willing to convey some sense of their experience of being in this state, instead of feeling shame about the absence of prior analysis on their part. This is not a straightforward choice between unthinking action and critical evaluation – people can often carve out a route between these two states in their conversations with us. Turning to a different interpretation of what is going on when people speak of doing 'work', Nicolini (2009) sees this as frustrating for researchers interested in professional practices because it indicates that people are unwilling to go through the procedural detail that he particularly prizes in his organisational studies. But we could also see it as an insight. In other words, whether people are excited by going through that detail (or prefer to sidestep it by deploying the idea of 'just being at work' to indicate that further enquiry is unwelcome) tells us something about how interested they are in exploring certain topics. We could see the idea of 'work' as a kind of conversational resource – an interactional template with which to control the spoken exchange, to discourage further examination and to push the discussion towards more appealing matters. Our spoken representations have a kind of 'force' (Anderson 2019) in the sense that they do things – they are interventions into the situation at hand, rather than straightforward depictions of the thoughts and ideas that are assumed to predate the occasion. Schmidt similarly points to how the practice of football has evolved over time partly through spoken interventions from those associated with the game (2017). Spoken exchange is not always about straightforwardly sharing our ideas, or about carefully coaxing out the truth of how people apparently 'see' the world. It could equally be about how certain ways of speaking about their practices may have practical impacts such that people might handle such an exercise with care. When Wacquant's boxers refuse to talk too much about the detail of how they box this is a deliberate shying away from something they have good reason to leave untroubled by reflection. We can learn from these reactions.

3. Practice talk activates norm ideas

Whilst my presumption was that there may be many ways of staging useful talk about the social practices that live through people (and that doing so might not always be as onerous for them or challenging for us as we might have otherwise thought), it is also true that some of the power of social practices stems from their assumed 'social' quality. Put simply, part of the reason why we so willingly submit to them is because, as part of that process, we often develop a background sense that, in so doing, our actions are probably shared (that we are responding to the situation in a socially agreed upon way). In other words, practices can become stronger when the individuals involved start to think of their actions as 'the done thing' since, once certain actions are thought of as such, they no longer invite any critical inspection (Nicolini 2017; Schmidt 2017). But perhaps the idea of 'the done thing' is better imagined as a notion that is invoked in the heat of the moment to justify actions that may not always be all that similar to those of others. Perhaps we make appeals to 'common sense' in the situation (rather than having a strongly ingrained sense of certain actions being all that 'common'). When Browne (2016), for example, asked groups of peers to talk together about their domestic water consumption, ideas to do with the 'done thing' were drawn on as another conversational resource – a suggestion that was invoked Jefferson under the circumstances of being asked. In so doing, her respondents also ended up having a lot of fun exploring how and whether they really did the same thing with their laundry (which also, to return, in passing, to the above discussion, rather goes against the idea of this necessarily being a stressful experience). Rather, and in line with what others have said about conversational laughter (Jefferson, Sacks, and Schegloff 1987), talking about practices can be an effective route to desired levels of intimacy. Both responses – 'activating' the idea that others are doing the same as us and enjoying the experience of exploring why that isn't always the case – might be particularly likely when people are asked to speak of their routines 'behind closed doors' at home (Shove 2003). In a study of domestic heating in older person households, we found that ideas related to 'norms' were carefully deployed when talking about what others of a similar age were doing. The aim was to acknowledge diverse circumstances whilst also positioning their peers as alike in doing what was 'sensible' for them (Hitchings and Day 2011). Indeed, it sometimes made for an enjoyable kind of gossip as these differences (that were not otherwise often spoken about) were raked over together with us. So, whilst we should be sensitive about how we study these matters, under the circumstances of being asked about everyday practices, notions of social norms can also come forth in revealing ways. There are different ways of thinking about how we handle 'the done thing' in our spoken exchanges and lessons to be learnt from how and when this idea is deployed by people. In that sense, we might think about how notions of shared collective 'culture' operate as a kind of 'toolkit' that helps people to justify certain actions (Swidler 1986). Again, we see this in

Wacquant when he explores the 'right way' to box and the idea of what a 'serious boxer' should rightly be. He learnt about these norms partly from a locker room tirade from a 'rising star' at his gym. Here the boxer involved emphasises the many rewards that await them, were his peers only to commit themselves fully to the practice (Wacquant 2004, p. 50). Under the conditions of being frustrated by their lack of enthusiasm, he activates this idea in pursuit of a stronger shared sense of the 'done thing'.

Is a 'Practice Interview' Possible?

So far, I have been quite careful about how I characterised what happened in the four studies on which the book draws. I've generally talked about 'speaking with people'. But in what ways? Where? And how? And should these encounters be correctly thought of as 'research interviews'? I turn to these questions next because, as I discuss, the ways in which we imagine the social science 'interview' can have important implications.

A question that I've repeatedly asked myself when writing this book is whether I did 'interview research'. On the one hand, I clearly did. In all of the projects discussed in this book, I arranged to speak with identified individuals in advance. They were defined as those belonging to particular groups that were of interest to me. I briefed them before we met. I said things about being anonymous and about how they should stop our discussion at any point if they felt uncomfortable. I said some things about what I was planning to do with my recordings afterwards so that placing a Dictaphone in front of them didn't prove to be too unnerving. I also revealed a little about what I was aiming to explore in my studies and discussions, though I usually did this in relatively vague terms because I wanted to see how they would handle talking about their lives in the moment (instead of deciding upfront how they wanted to present themselves and then defending that position). Generally, I had some sheets of paper in front of me upon which I had typed out some of the questions and topics that I hoped to cover over the course of our time together. For some of the discussions that happened outside, I even turned to the trusty clipboard to protect my sheets from the elements so that I stayed on track and covered all my questions despite the effects of weather (the pesky outdoors!). During our time together (sometimes out of habit and sometimes out of nervousness), I would use my pen to strike through my topics and questions as the discussion went on so that, by the end of our time together (usually an hour), I was fairly confident that I had covered them all. All this was very much along the lines of a traditional 'social science interview'. Indeed, I did several things because they were 'the received wisdom' about effective research interviewing (such that I dutifully followed suit). This approach was what respondents seemed to expect too – it reassured them I was doing my job properly, that I was appropriately serious about my studies.

On the other hand, it was often much less of a fixed procedure than that. If you consult books on social research methods, the interview is often imagined as two people facing one another across a table in an otherwise unremarkable room (the less remarkable the better). The idealised interaction is, without doubt, a focused form of exchange – there is a job to be done here and the researcher has ideally arranged things so that the views of the person being spoken with are extracted and explored in the most efficient way possible. The talk that is analysed in my case studies occasionally happened like that. But this was far from the rule. Some conversations took place as we sat together drinking wine and eating crisps, mulling over the domestic garden that we could see through the patio doors and wondering about how it had gradually come to assume its current form. Some took place in busy city restaurants as some of the professional office workers slotted a burst of speedy talk with me into their otherwise time-pressured lives in ways that were already telling me something important. Some began immediately after we stepped off adjacent treadmills as I struggled both for breath and for the right way to start speaking about what had just happened. And some suddenly started after quickly sitting down on the grass because I felt I should immediately capitalise on the passing interest of those who had been caught on the hoof at a music festival. These situations led to all sorts of exchanges.

My point here is not about context, namely to say (as many have) that where we do our interviewing influences what gets said (though, as we will see, that definitely played into all the projects discussed here). Rather it is to emphasise how I adapted my interviews to the social situation (and how it made sense to talk with those involved at that time). I say this because one problem that I see with some of the above accounts is that the interview features within them as quite a fixed thing – a straightforward extractive exercise. In some of the accounts, interviews are pilloried partly because interviews are imagined to have characteristics and ambitions that are antithetical to the proper investigation of everyday life. Interviews can be seen as problematic in this field of work because they take respondents out of the current of activity in which they are generally involved. The idea, in other words, is that they cannot help but pull people out of their practices in a way that leads to an unsatisfyingly anaemic account of what happened before. This is either because people cannot talk about the detail or because they don't particularly want to discuss it.

If we thought of our spoken exchanges in a less prescribed way, we could see them in a more positive light. Accordingly, in terms of the interviews on which this book will focus, I see them as a collective endeavour in which we came to a shared appreciation of how certain aspects of people's lives were organised and experienced. And, in terms of how I 'took an interest' in the everyday practices of my respondents, I asked a variety of questions according to what seemed to yield the most interesting answers and how they provided different forms of access to their experience. I didn't want to recreate 'the conditions of sterility' that are imagined to foster the efficient extraction of people's perspectives and which are sometimes associated with the ideal social science interview (Deutscher, Pestello, and Pestello 1992, p. 93). Rather the aim was to pay particular attention to the 'conversational footings' (Potter and Hepburn 2005, p. 293) involved, namely how certain questions and interactions helped me to understand their lives. In this sense, what I did was in line with the idea of the 'active' research interview – one in which we considered particular topics together and in which I took on particular roles (just as they did) according to what proved instructive for me and comfortable for them in the moment (Gubrium and Holstein 1997). Returning briefly to my sheets of paper (that I'd gradually scribble over in order to know when each question had, more or less, been covered), as I struck through these questions, people sometimes reacted to a process that I assumed they wouldn't notice. I think they sometimes thought that I was judging them (Wait, why are you not asking me that question? Why have you crossed that one off your list? What are you concluding about me?). Learning from this kind of response was only possible once I allowed myself to think of my (what I'm now happy to say were) 'interviews' as rather different to the collection of pre-existing respondent perspectives in appropriately neutral scientific conditions.[1]

Taking the Piss and Holding Your Nerve

Just as how it is useful to think through the implications of how the social science interview is commonly imagined, so it is useful to think about what is frequently said about the interactions that should happen within it. Valentine (1997), for example, suggests that we should think of the ideal interview as a dialogue more than an interrogation. McCracken (1998) recommends that the interviewer should be benign and accepting, curious more than inquisitive. Rapport seems paramount here and that obviously makes a lot of sense. Without effective rapport, after all, the whole enterprise could very easily collapse as conversations are shut down by disgruntled respondents who clam up and give cursory answers that don't tell us so much and leave little time to identify effective follow-ups.

In this respect, though I've presented it as a potentially enjoyable activity for both parties, this is not to say that 'taking an interest in the everyday lives of others' was always easy. There were instances when what I was doing made me worry that rapport was indeed under threat and the prospect of conversational collapse began to loom large. Taking an interest in the everyday lives of others, perhaps more than other topics, required me in these moments to stay determinedly focused on my interviewing ambition of exploring what people had to say about mundane matters that might easily be dismissed as unworthy of comment in other contexts. Certainly some of the topics that I wanted to consider will have seemed strange to discuss with a comparative stranger.

In the United Kingdom, people sometimes accuse each other of 'taking the piss'. This disparaging phrase can be used to imply that your conversational partner is feigning interest in a topic that they actually think is unimportant and uninteresting. Sometimes I felt that my respondents were momentarily worried about me potentially 'taking the piss' in exactly this sort of way. There were definitely occasions when I observed a flicker of judgement flash across their faces as they held my gaze for a second to see whether I would crumble under this scrutiny and belie the truth – that, in actuality, I found these discussions rather silly. Perhaps they were hoping for a conspiratorial smile or snicker from me to diffuse the situational worry. Some of them were, I think, briefly anxious that I was daring them to talk about inconsequential matters before going away to mock them elsewhere about their willingness to do so.

I'd try to 'hold my nerve' in these situations. And, in so doing, our spoken exchanges eventually settled into a collective appraisal of how aspects of their lives that had hitherto not often been discussed (and certainly not with someone who had just turned up) went on. This is not, however, to say there was no fun to be had in embarking on this kind of exchange. But it was more that the humour came from them in terms of how they wanted to present their observations to me – when they found it comical that that were doing certain things that perhaps, on reflection, didn't make too much sense, or that they were thinking (or, just as often, not thinking) in certain ways about various actions that were familiar to them but went unexamined before. The point was to attend to how they managed our talk about these matters. Sometimes, but not always, people were controlling the conversation because they wanted to paint their actions in particular ways. And that was often both to themselves and to me. Sometimes, but not always, they seemed to be approaching annoyance in response to the suggestion of incorrect conduct on their part. But that was often a suggestion of their own making rather than my own. Sometimes, but not always, in our interviews they dragged up a justification for something they never usually considered. But it was often openly acknowledged that this was what was happening. Together we poked and prodded at their everyday practices. And, through these means, I gradually came to understand their relationships with them. What follows is the result.

Back to the Outdoors

In the last chapter, I made a case for looking at how certain common social practices might play an important part in keeping people away from positive encounters with greenspace. So far in this one, I have considered how we might study how these practices capture people. More specifically, I have thought about how we might go about undertaking spoken exchanges with those who carry out certain practices and I have lifted the lid a little on some of my own attempts at doing so. My focus was established in the last chapter. In this one, I've now identified a number of suggestions about how I might explore it. I already had a topic. And now I've developed an approach. But where should I go to test it out?

Next the book turns to four case studies. Each case centres on a social group who might, by virtue of their ongoing relationship with a particular practice, find experiences with the outdoor environment challenging (or rather that certain experiences were effectively being repelled by the practice they have come to carry). All four practices are widespread in the United Kingdom. For some, you might be able to easily imagine how they are likely to (sometimes insidiously) influence the ongoing relationship between groups of people and outdoor environments. Others might be less obvious in terms of immediately seeing the connection.

Why were these chosen? Though there could be many cases with which I could explore the potential of the book's approach, these were selected because the practices involved were common (both in my country and around the world) and because (according to my hunch at least) they could be distancing those involved from certain outdoor experiences. Returning to the arguments made by Bixler and Floyd (1997) that were discussed in the last chapter, changing patterns of everyday life could (in ways that those involved may not particularly notice) be serving to render certain aspects of nature 'scary, disgusting and uncomfortable'.

But what practices?

- I start with a context in which we find many people, namely the office. The aim is, first, to consider how the suggestion of neutral, climatically controlled and environmentally disconnected rooms have become seen as a desirable place in which to put employees: how the idea of sealing office workers inside spaces that may very well render them comparatively oblivious to the local outdoors has taken hold. Then I'll consider the ways in which such workers have been studied in terms of their relationship with beneficial greenery and how the present approach might usefully augment this work. As with all my studies, I then introduce the group with whom I spoke in my own study. In this case, it involved conversations with city lawyers at different points in their career. Then I consider what the detail of our interview talk tells me about whether and how these professionals might venture out into the greenspaces found outside their climatically controlled offices. So, the first practice on which I train my attention is office work. It is one that has, according to some, effectively become hidden in plain sight. In other words, it is exactly because so many of us do it that less has been said about the details of how office work is lived out by people. Yet that is one good reason why it is worth studying.
- The second case turns to the imperative to undertake regular exercise, as a societal injunction that is taken to be of growing importance (given growing anxieties about the public health implications of increasingly sedentary people). Exercise sits in an intriguing position between work and leisure. We might imagine there is more time to engage with environments outside

during exercise. But there is still often a job to be done here and there are objectives to be achieved. This is not play. It is with that tension in mind that this chapter considers where exercise currently happens with reference to the practice of recreational running. Like office work (though many office workers would unlikely frame it as such), recreational running has become increasingly popular in countries like the United Kingdom. However, we know less about how and why it has become attached to some environments instead of others (and how it can be examined with reference to the rise of the indoor gym). With that in mind, this chapter considers why some London runners have come to run on treadmills indoors whilst others are found outdoors on pavements and paths. The discussion starts with the popularity of gyms and how, like the offices of the first case study, they present themselves as attractive partly by virtue of marking themselves off as spaces apart from the outdoor environment or, in other words, as exclusively indoor spaces. Then we turn to the experience, in this case of treadmill running. My interest here is in how certain patterns of talk and thought may serve to keep some Londoners on the indoor treadmill, irrespective of where, on reflection, they'd ideally like to run.

- The third case turns to the management of outdoor spaces at home, more specifically the matter of what to do with a domestic garden. The home is often thought of as a haven of relaxation away from work. Maybe it is therefore in the garden that certain features of the outdoors are most welcome. Chapter 5 considers this suggestion by exploring how living plants fare against certain harder, more docile materials with which they compete for space in the garden. It begins with what others have said about the benefits of 'therapeutic horticulture'. Then it considers how those who sell the idea of a living garden must contend with the suggestion that many owners now have less time for tending and might rather prefer to efficiently create aesthetic impact by getting their gardens 'done'. With that in mind, the practice that provides me with an analytical anchor is that of shopping. This one in which many members of society are increasingly skilled, just as many are decreasingly skilled in gardening. With that in mind, I turn to how some of those who own a domestic garden in London speak with the 'garden designers' that they might hire. Engaging with living plants can, as this third study found, often require a repertoire of talk and thought that many of these garden owners were yet to possess. This chapter explores the challenges of developing one and draws out the implications for the future of beneficial greenspace experience in gardens.

- The final case ends the discussion of my fieldwork with a temporary escape from the city in which all three previous studies took place. The idea is that the disruption associated with a spell away from the city might tell us something revealing about how everyday practices are sustained back home. The focus is on what follows when young people are faced with certain hitherto unfamiliar ways of 'simple' outdoor living at the summer music festival. Can this experience be thought of as enjoyable and how easily do certain strategies for coping with life outdoors spread through the festival? As with the preceding three cases, the discussion begins with existing research on relevant greenspace benefits (in this case how camping has been considered). Then it turns to how those hoping to profit from providing 'outdoor hospitality' are responding to the perceived desires of a new group of customers who are unwilling to take on the challenge of the outdoors too fully. My final practice is then introduced and it is here that the link to 'everyday life' is made. Attention now turns to regular showering as an activity that has become particularly popular amongst the young people who also predominate at the music festival. This leads to questions about how this practice responds after they set up camp far away from the showers that provide a background support in their everyday lives back home. How easily do the standards of those who somehow feel compelled to uphold them elsewhere slip at the festival? And what is revealed about this process by attending to the patterns of talk that both circulate through festival spaces and can be observed within research interviews? In this way, though away from the city, the festival acts as an 'experimental space' from which we can learn about how attached relevant groups are to certain forms of bodily management back in urban life (and how welcoming they are to the potential disruptions of outdoor experience).

Insights from the Edge

Many of those with whom I spoke in these four studies were picked because of the strength of their relationship with the practice at hand (those who spend a lot of time in offices; those who generally run indoors; those who are quite skilled at shopping; those who wash more often than most). If we understand these practices as potentially spreading, they were the potential pioneers found at the edge of social change. In this regard, this book aims to provide an account of how it is to live at this edge, along with the likelihood of those involved taking future paths that lead both towards and away from certain beneficial greenspace experiences. In that sense, my aim is to provide insights from the edge.

Were you to paint a picture of someone who might be involved in all four studies, however, the mental image you might begin with is of quite a privileged person. I studied those who work in comparatively plush offices with controlled ambient conditions and on-site facilities. I studied those in a position to pay for the opportunity to exercise inside private gyms instead of having to risk more public environments outdoors. I studied those who are lucky enough to be able to pay for certain types of domestic gardens instead of having no choice but to find a way of managing it themselves. And going to a music festival in the United

Kingdom, though an increasingly common experience, can be expensive. In one sense, it would be fair to characterise this as a book on the pampering of the contemporary middle classes.

But to do so would be to return to an image of society as comprising social groups when part of the point of starting with social practices is to see it differently. And so, whilst there will be race, gender and class differences that will all likely play into the processes on which I am focused, the point is not to start with these comparatively familiar distinctions. Instead, the aim is to see society as comprising practices, some of which could easily recruit many more carriers, depending in part on whether they have the financial, physical and circumstantial capacity to be recruited. And so, whilst I would welcome studies that look at how these other dimensions kick in elsewhere, that's not my focus now. My aim is to explore the experiences of those who were chosen because of how they were potentially pioneering a broader process of environmental estrangement that was yet to receive much attention.

London as the Laboratory

Before we get stuck into my case studies, I should say something at this point about the city in which I did the majority of the fieldwork for this book. First of all, it is worth noting how London is a city rich in greenspace (see Figure 2.1). Greenspace in London is equivalent to 40% of the total surface area – making London one of the greenest cities of its size (BOP Consulting 2013). So, for those who, like me, are keen on thinking about greenspace benefits in an urbanising world, London may seem like the perfect place to find positive stories. There has even, for example, been an ongoing discussion here about whether London should be designated a 'national park'.[2] Though it may seem counter-intuitive to think of a city in this way, that is precisely the point: London is so well stocked with greenspaces that it may nonetheless warrant such a designation.

But, as was argued in Chapter 1, we should be careful not to jump the gun by focusing too soon on provision since the 'opportunity' to engage with such spaces may often be trumped by the 'orientation' towards the idea of going into them. So, whilst it may be tempting to feel positive about maps such as shown in Figure 2.1, other depictions would tell a very different story. What if we plotted the spread of indoor gyms across the city that may pull exercisers indoors and away from these greenspaces? What if our maps sought to depict how concrete and paving has been squeezing plants and trees out of front gardens in ways that mean people see less greenery as they sit inside their homes? What if we mapped how air-conditioning has spread through the city and consider how indoor climate control can make people disinclined to take on the challenges of being outdoors? There are many

0   2   4   8   12   16
Miles

Figure 2.1   Areas defined as publicly accessible green space in Greater London. Source: Vivid Economics 2017.

stories waiting to be told about the changing relationship between people and greenspace in cities like London and, if we focus too soon on provision, we will only ever be able to learn from some of them.

On the matter of context, and on the status of London, there are many features specific to London that have some bearing on the processes discussed in the following chapters. Beyond its greenspaces, we can think of how the mild local climate and the variable weather might act to make the outdoors both attractive and challenging for Londoners. Then there are the pollution levels (higher here than in some cities, but lower than in others) that may also encourage particular ways of feeling and acting when people venture outside. There is also the social experience of public life outdoors in cities and how, though London has its moments, there are still many spaces there where it is possible to pause and look beyond the distractions of others around you. Then, if we think of some of the practices on which the book focuses, the 'long hours' working culture of the United Kingdom is likely to kick in too. So, for those who may be wondering why London was chosen (and whether the processes described in this book should be thought of as specific to this city), my answer is that there will definitely be variations that make it worth considering how and why the situation is different in Singapore or Sydney, in Beijing or Abu Dhabi. For now, the point is rather to explore the promise of the present approach and to use that to develop some suggestions that might then be tested out elsewhere. London is my laboratory (and it was also, admittedly, partly chosen because it is where I live and work myself). In other words, it provides an opportunity to develop the present approach and then see where that leads.

… and Back to the (Apparently) Banal

With that said, I now turn to my case studies. In each of them, I provide an account of how a widespread social practice is shaping relationships with outdoor greenspace and how I examined that process by talking with some of the people involved. In each of them, I'll also identify a way of characterising the process at hand and consider the relationship between my interview experiences and the three suggestions developed above. The final chapter starts by comparing the findings of the four case studies before taking stock of what has been learnt from this exercise (in terms of both greenspace benefits and practice interviews). But I don't want to pre-empt my findings at this stage. The point is rather to learn from taking an interest in the everyday lives of those who I found in various places (and seeing what comes from speaking with them about matters that may initially seem unexcitingly banal). I'll begin with everyday life inside the corporate office.

Endnotes

1 None of this is to suggest that I am the first to argue for the value of interviews in this field. For example, though Bourdieu suggested that routine practices are 'opaque' (1990, p. 12) to their practitioners, he also goes on to advocate a certain kind of 'disposition analysis' where the aim is to understand how it feels to embody a particular practice (Everett 2002, p. 7). This is commonly done by a form of 'accompanied self-analysis' (Bourdieu 1999) where our academic tendency to examine the situation within which we find ourselves (Sweetman 2003) is shared with our subjects as a means of helping them to evaluate their own lives. It's more that I want to take this opportunity to use my four studies to develop interview approaches to this topic (Hitchings 2012). I feel that I have now probably done enough relevant talking with various groups of people to be in a fairly good position to do so. After all, whilst novel theoretical approaches can sometimes be taken to demand methodological innovation (Hitchings and Latham 2020), it is definitely also worth exploring how existing methods can be used to explore alternative ways of imagining everyday life.

2 See, for example, the 'National Park City' movement, the aim of which is to make London simultaneously 'greener, healthier and wilder' (http://www.nationalparkcity.london accessed 27 January 2021).

References

Anderson, B. (2019). Cultural geography II: the force of representations. *Progress in Human Geography* 43: 1120–1132.

Becker, H. (1998). *Tricks of the Trade: How to Think about Your Research While You're Doing It*. Chicago, IL: University of Chicago Press.

Bissell, D. (2014). How environments speak: everyday mobilities, impersonal speech and the geographies of commentary. *Social and Cultural Geography* 16: 146–164.

Bixler, R. and Floyd, M. (1997). Nature is scary, disgusting and uncomfortable. *Environment and Behavior* 29: 443–467.

BOP Consulting. (2013). Greenspace: the benefits for London. Report prepared for the City of London Corporation. London: Author.

Bourdieu, P. (1990). *In Other Words*. Stanford, CA: Stanford University Press.

Bourdieu, P. (1999). Understanding. In: *The Weight of the World: Social Suffering in Contemporary Society* (ed. P. Bourdieu and P. Parkhurst Ferguson), 607–626. Cambridge: Polity Press.

Browne, A. (2016). Can people talk together about their practices? Focus groups, humour and the sensitive dynamics of everyday life. *Area* 48: 198–205.

Crossley, N. (2006). In the gym: motives, meaning and moral careers. *Body and Society* 12: 23–50.

Deutscher, I., Pestello, F., and Pestello, F. (1992). *Sentiments and Acts*. New York: Aldine de Gruyter.

Everett, J. (2002). Organisational research and the praxeology of Pierre Bourdieu. *Organizational Research Methods* 5: 56–80.

Giddens, A. (1986). *The Constitution of Society*. Berkeley, CA: University of California Press.

Gubrium, J. and Holstein, A. (1997). Active interviewing. In: *Qualitative Research: Theory, Method and Practice* (ed. D. Silverman), 113–129. London: Sage.

Halkier, B. (2017). Questioning the 'gold standard' thinking in qualitative methods from a practice theoretical perspective: towards methodological multiplicity. In: *Methodological Reflexions on Practice-Oriented Theories* (ed. M. Jonas, B. Littig, and A. Wroblewski), 193–204. New York: Springer.

Hitchings, R. (2012). People can talk about their practices. *Area* 44: 61–67.

Hitchings, R. and Day, R. (2011). How older people relate to the private winter warmth practices of their peers and why we should be interested. *Environment and Planning A* 43: 2457–2467.

Hitchings, R. and Latham, A. (2020). Qualitative methods III: on different ways of describing our work. *Progress in Human Geography*. https://doi.org/10.1177/0309132520901753 (accessed 27 January 2021).

Hui, A., Schatzki, T., and Shove, E. (2017). *The Nexus of Practices: Connections, Constellations, and Practitioners*. London: Routledge.

Jefferson, G., Sacks, H., and Schegloff, E. (1987). Notes on laughter in the pursuit of intimacy. In: *Talk and Social Organisation* (ed. J. Graham Button and R. Lee), 152–205. Clevedon: Multilingual Matters.

McCracken, G. (1998). *The Long Interview*. London: Sage.

Nicolini, D. (2009). Articulating practice through the interview to the double. *Management Learning* 40: 195–212.

Nicolini, D. (2012). *Practice Theory, Work, and Organization: An Introduction*. Oxford: Oxford University Press.

Nicolini, D. (2017). Practice theory as a package of theory, method and vocabulary: affordances and limitations. In: *Methodological Reflexions on Practice-Oriented Theories* (ed. M. Jonas, B. Littig, and A. Wroblewski), 19–34. New York: Springer.

Pichelstorfer, A. (2017). (Re)configuring actors in practice. In: *Methodological Reflexions on Practice-Oriented Theories* (ed. M. Jonas, B. Littig, and A. Wroblewski), 79–92. London: Springer International.

Potter, J. and Hepburn, A. (2005). Qualitative interviews in psychology: problems and possibilities. *Qualitative Research in Psychology* 2: 281–307.

Schatzki, T. (2017). Sayings, texts and discursive formations. In: *The Nexus of Practices: Connections, Constellations, and Practitioners* (ed. A. Hui, T. Schatzki, and E. Shove), 126–140. London: Routledge.

Schmidt, R. (2017). Reflexive knowledge in practices. In: *The Nexus of Practices: Connections, Constellations, and Practitioners* (ed. A. Hui, T. Schatzki, and E. Shove), 141–154. London: Routledge.

Shove, E. (2003). *Comfort, Cleanliness and Convenience: The Social Organisation of Normality*. Oxford: Berg.

Shove, E., Pantzar, M., and Watson, M. (2012). *The Dynamics of Social Practice*. London: Sage.

Sweetman, P. (2003). Twenty-first century dis-ease? Habitual reflexivity or the reflexive habitus. *The Sociological Review* 51: 528–549.

Swidler, A. (1986). Culture in action: symbols and strategies. *American Sociological Review* 51: 273–286.

Valentine, G. (1997). Tell me about using interviews as a research methodology. In: *Methods in Human Geography: A Guide for Students Doing a Research Project* (ed. E. Flowerdew and D. Martin), 110–253. London: Longman.

Vivid Economics. (2017). Natural capital accounts for public green space in London: report prepared for Greater London Authority, National Trust and Heritage Lottery Fund. London: Vivid Economics.

Wacquant, L. (2004). *Body and Soul*. Oxford: Oxford University Press.

Chapter Three
Forgetting the Outdoors: Inside the Office

Making the Floating Factory

Office work may initially strike you as an uninspiring topic for sustained analysis. Sitting at a desk, writing emails, doing various forms of paperwork, occasionally going to meetings in another unremarkably beige room: these are some of the activities that we might picture if asked to think about office work, and they probably don't pique our interest all that much. Yet, one of the arguments underpinning the project that provides the focus for my first case study is that is precisely why they should. Around 70% of those working in advanced industrialised countries are now involved in 'information work' (de Decker 2016) and, though things have been getting more mobile (Hampton 2017), that still generally involves a good amount of time spent in offices. Indeed, it is precisely because office work has become so ubiquitous that it has become so unremarkable – the kind of activity that falls away into the background of our consciousness. But that is also what makes it worth studying since so many people are now doing it. As a neat strategy for encouraging us to recognise this, Van-Meel (2000) asks us to take a moment to consider the contemporary skyline of most cities. If we did, he argues, we would be immediately jarred out of this indifference. Why? Because what often dominates these scenes are forests of, often speculatively built, glass

The Unsettling Outdoors: Environmental Estrangement in Everyday Life, First Edition. Russell Hitchings.
© 2021 Royal Geographical Society (with the Institute of British Geographers). Published 2021 by John Wiley & Sons Ltd.

office blocks. This chapter peers inside a sample of these blocks to explore how the inhabitants live and work inside and what this reveals about their relationship with the outdoors.

But before we get to that, some words on how a situation that we could easily argue is strange (and potentially unhealthy) came about. The idea of the modern 'office' emerged at the turn of the twentieth century as a strategy for efficiently organising workers, often women (Strom 1994), to perform administrative tasks within a defined physical space. The idea of 'scientific management' was a key idea in formalising how 'office work' was to be organised as engineers, as a set of people with a particular way of responding to problems, applied their skills to this challenge (Baldry, Bain, and Taylor 1998). Those found in offices were initially subject to the same kind of scrutiny as those who laboured in factories as these engineers focused on how space could be most effectively organised so that people could complete sequences of tasks most productively. In this way, the environment came to define the practice. There are not all that many jobs that are so strongly linked to where the activity happens. When teaching my students, I don't call myself a lecture-theatre worker.

But why seal these workers inside in the first place? The most common response to this question is to tell a story about the efficient use of valuable urban space. The buildings in which offices are frequently found have undergone many changes over the years as various technological developments and architectural trends filtered through into different visions of what offices should look like and what those working inside them were taken to require. For example, office build-ings were initially designed to make the most of light from outside by having 'T', 'L' or 'H' shaped cross-sections (de Decker 2016). This meant that no individual office was too far away from the outside environment. However, with the arrival of air-conditioning and artificial lighting, this situation soon started to change. After this point, work was increasingly taken to be possible further away from the building's edge (by supplying inhabitants in the heart with artificially con-trolled forms of air and light). As a result, buildings could become boxier and office workers could be pushed further inside. They were also pulled up and away from the ground through the institutionalisation of elevators that helped them get quickly to their offices and get on with their work. All this resulted in the gradual spread and current predominance of the office 'block' format.

Generally speaking, these office workers were thought to benefit from an unre-markable experience (no surprise then that these lifestyles often escape attention today if they were designed to go unnoticed by those doing this work right from the start). In terms of air-conditioning, for example, engineers of 'thermal comfort' encouraged building managers to provide occupants with specific conditions that meant they no longer needed to think about how best to regulate their bodies in the workplace (Shove 2003). But specifying conditions could also decouple workers from the outside world in certain insidious ways. We can, for example, think about how a strategy underpinned by the seemingly benign aim of providing worker 'comfort' served to underwrite the global spread of the standard 'business suit' as appropriate professional attire. Though the intention was only to supply the temperatures that professionals wanted, by providing the same thing everywhere, they ended up wearing particular kinds of clothing that were often ill suited to the local climate outdoors. And if you are provided with a cool environment throughout the day, and you are expected to dress in a way that depends upon cool temperatures and certain humidity levels, you might soon come to shun outdoor environments because they are so different to those to which you have adapted indoors. If asked to imagine a figure who lives in line with the environmental decoupling that Ingold (2004) worries has made the experience of contemporary urban life so uninspiring, it could well be the professional office worker who springs to mind.

Yet there could be hidden anxieties harboured by those who labour within these seemingly anodyne spaces. We could recall the sick building syndrome scare in which office workers gradually forged a social movement from what were merely passing discomforts at the start (Murphy 2006). Identifying how to respond to them would, however, require some care; for example, when office workers are given control over their conditions, this doesn't always lead to more satisfaction or productivity (Lee and Brand 2005). Perhaps this is less about environmental control and more about environmental connection? Either way, given current worries about increasing numbers of office workers being stressed (and concerns about how working environments could very well be part of the problem), perhaps it is the common office worker that we should consider in terms of how nearby plants and trees could help (Gladwell and Brown 2016)? This suggestion has provided the motivation for a group of scholars who have explored the relationship between office workers and greenspaces. And so, before turning to the group whose accounts provided me with the material for my own analysis, it is worth considering how others have tackled this topic.

Plants Coming In, People Going Out

A range of researchers and designers have sought to investigate whether and how 'nature' (which they most often define as areas of greenery of various sizes – from plants inside to landscaping outside) can be used to benefit the great number of people found inside offices. Going beyond issues of workplace stress, the potential benefits of being near to nature in the workplace have been understood in various ways (from a strategy for dampening angry feelings to a means of encouraging job satisfaction). Drawing on these studies, some have emphasised the importance of getting the relationship between office workers and this particular kind of 'nature' right (Kaplan 1993). In so doing, the rationale moves from a benevolent public health interest to a business case about productivity. Fostering greenspace benefits at work could be good for all sorts of groups, after all.

In view of this agenda, it is unsurprising to see a focus on helping office workers to experience plants visually. Here the argument goes that, by making sure they can see vegetation from their offices, they may benefit from 'micro-restorative experiences' (Kaplan 1993) – short bursts of looking that take them away from their immediate tasks (with the added employer bonus of no-one leaving their desks). Then they'll likely return to what they were doing feeling refreshed. In light of findings that suggest the value of these breaks, some have even argued for redesigning cities with this as a guiding principle. The thinking here is that, if we spend more time at our desks than we do going to and from work, then encouraging office workers out into greenspace should not really be the aim. Rather, the outdoor spaces that we already have in our commercial centres should be filled with trees that can be seen by the maximum number of office workers from the glass boxes they may be disinclined to leave during the day. Perhaps we should invest in elevated courtyard gardens that are closer to those with offices higher up. Maybe we need window boxes that can be seen by those working on each level of the block (Chang and Chen 2005). If we think it is too much to ask people to venture down and outside, this makes a lot of sense. Perhaps we should consider issues like how providing greenspaces that seem 'wilder' outside may take people more effectively away from their tasks (though it also seems that many office workers like to see evidence of appropriate levels of control and maintenance when they look down upon areas of outdoor greenspace) (Kaplan, 2007; Colley, Brown, and Montarzino 2016).

If the assumption is that office workers are either unwilling or unable to leave their desks, another route to restoration would be to take the greenery to them. Perhaps it would be better to provide them with plants (or even pictures of plants and trees) in their offices (see, for example, Larsen et al. 1998; Shibata and Suzuki 2002; Kweon et al. 2008; Thomsen, Sønderstrup-Andersen, and Muller 2011). Nearby plants can seemingly help people to feel more positive about what they are doing at work, though (and going back to the employer benefits), this may not always lead to additional productivity. Perhaps when plants draw attention away from work it is not always so easy to return. Perhaps they can also physically get in the way if people are not actually so chained to their desks. Even pictures of natural scenes can seemingly help. Indeed, it appears that office workers are often well aware of that. This, at least, is what is suggested by the 'compensation hypothesis'. The idea here is that, when office workers are forced to forgo the opportunity to see actual vegetation by looking out from their windows, they'll likely take steps to address this situation by putting up pictures that depict natural scenes near their desks (Heerwagen and Orians 1986; Bringslimark, Hartig, and Gridal Patil 2011). These options might be especially worth considering if the reality of life inside offices means that it is increasingly unlikely that workers will endeavour to experience the 'real thing' outside.

But this is far from a done deal. Indeed, it is with that in mind that a smaller number of studies have considered the likelihood and effects of these workers venturing outside to benefit from more direct greenspace experience (Hartig

2006; Gladwell and Brown 2016). Some suggest that, if they did, this could lead to additional benefits (since doing so means they are not only getting some restoration from being around plants and trees but they are also getting a small amount of exercise). This is valuable when the trend towards office work is often seen as one of the chief adversaries in the public health fight against increasingly sedentary societies (in itself perhaps no surprise in view of all the effort that was put into keeping office workers away from any annoying environmental distractions). Here one study suggests that the length of time outside, more than the frequency of going, may bring the most mental benefit (Gilchrist, Brown, and Montarzino 2015). Perhaps it takes time for the 'otherwise preoccupied office worker' to reach a point when they really get something out of it (unlike the idea of the 'micro restorative' experience)? Equally, this same study finds that workers who are more, rather than less, stressed are those who are most likely to be found in greenspaces around their offices. Maybe they only go there when they really need to. Perhaps nearby parks and gardens have become a kind of 'emergency resource', more than a routinely drawn upon source of temporary workplace respite. Others suggest that those within a corporate office context feel reassured by simply knowing that the greenspace is there if they need it (Lottrup, Grahn, and Stigsdotter 2013). They might be especially mindful of the relaxation risks of spending too much time outside (just as their employers might be). Though aware that going out into greenspace could provide an enjoyable and restorative experience, if workloads will likely build up whilst you are away, perhaps it is better not to go after all!

These studies hint at some of the practical challenges of encouraging a closer relationship between contemporary office workers and outdoor environments in everyday life. But what of the detail? Because of a preference for survey studies in this field, we currently know less about how different office workers manage their relationships with different forms of greenery. In one of these studies (Lottrup et al. 2012), a large sample of Danish office workers were asked why they didn't take more breaks outside. A large chunk of their respondents said there were 'no real impediments' for them. Few employers would, of course, be so heavy handed as to forbid it (stay at your desks, we've already provided some plants for you to look at from your workstation so you don't need to go out!). Rather the processes by which office workers do, or do not, 'connect' in some way with these spaces are probably more subtle than that. Indeed, they may be very much bound up with how the social and material environments of work are subjectively experienced and routinely handled. And that was what the first of my four case studies sought to explore.

A Group of Lawyers in London

Now we travel into the corporate heart of London to see what some of those working there have to say about their everyday lives and outdoor spaces. It's worth saying at the start that central London is blessed with a number of features that could make us feel quite positive about the potential for a healthy relationship between office workers and greenspaces. The City of London Corporation (2013), for example, manages over 200 'small' green spaces in the city and so, though they may be small, they are often nearby for many of these workers. However, these workers are also pretty packed in. Indeed, the same body commonly champions the number of office workers found in the city as part of an attempt to underline the economic value of this part of the city to the national economy. But the implication is also that these knowledge workers are crammed on top of one another inside their towering blocks in ways that mean that, were they all to go outside into the parks and gardens scattered around them, these spaces would soon get crowded. Furthermore, depending on how you look at it, they were also provided with 'facilities and services' that might insidiously serve to keep them inside – onsite gyms, subsidised canteens, even shops.

The people that I spoke with were all lawyers. I should therefore acknowledge up front that other professional groups will likely face different challenges. But there will also be ways in which their challenges are the same. In any case, as with all four of the studies in this book, the idea was as much to test out what could be learnt by applying the present approach as to say something more definitive about the nature of modern office work. Twelve lawyers were involved. Because I thought it would help if I got to know them a little (and because I wanted to see how, and whether, they responded to seasonal change in their everyday lives), I talked with each of them every three months over the course of one calendar year (so once per season). The justification for doing so was that, as one meeting followed the other, I would gradually develop a more subtle sense of how their working lives went on. It also gave me the chance to think about the best way of exploring my interests further with them based on what they said in our previous meetings.

Similar numbers of men and women were recruited since gender can influence the ways in which office workers are able to present themselves in bodily terms (and make for different kinds of workplace expectation). Respondents from seven firms were recruited and they worked in a variety of practice areas. Care was taken to ensure a mix of those who had recently started doing this kind of work and some comparative 'old hands'. In doing so, I wanted to see how relationships with the practice of office work may have changed for them according to their length of exposure to it. Four respondents had less than two years of experience as practising lawyers and the remaining eight were more fully established in this profession. Respondents were also drawn from different firms because different buildings can offer different opportunities with regard to outdoor experience and particular companies may have particular cultures with respect to the forms of everyday practice that they deem locally acceptable. Most came from the traditional heart of the city. Two worked in the newer financial centre of Canary Wharf. All, however, spent significant amounts of time labouring inside their offices. All of them were often found inside the floating factory.

Encouraging the Preoccupied to Pause and Reflect

These were busy people. And, in some respects, this made me worry that too much active examination of how it was to live in the ways that they currently did would be unwelcome. I also worried that, not only were they busy, they might also be disinclined to embark on certain discussions with me because they needed to keep going with their careers. As such, stopping to take stock of their situation might upend some of the contextual cues that they had either encouraged themselves, or with which they had been provided, that helped them to plough on through some pretty heavy workloads. But that was also (going back to some of the arguments from the last chapter) precisely the point – the aim was to see how they handled this kind of encounter and examine how exactly they spoke of life in the office. On that more positive note, I was also hoping that these features could also act to pique their interest. I was providing an opportunity to stop for a second and to consider their lives and, though time was often precious, that might be interesting to do.

Though there was plenty of rescheduling when work became too busy, no one dropped out. And so, once every season over the course of one year, I would pop up into their lives and we would talk about how office work and outdoor spaces currently did, and potentially might, connect for them. Sometimes we met in coffee shops not far from their workplaces. Sometimes I went to restaurants. Sometimes we met in meeting rooms since that meant they could return to their desks more quickly afterwards. Occasionally I went to meet them at their homes. There were definitely times when it was hard to encourage the discussion I was hoping for (to snap them out of their workplace dispositions in order to question how these dispositions had come about). On those occasions, I reminded myself that I was equally interested in what kept them within certain frames of mind (and what that meant for what they would and wouldn't do). When they were sometimes floored by a weirdly trivial question, that said something about how alien the idea was to their current practice.

In this way, and as discussed in Chapter 2, I needed to 'hold my nerve'. That didn't always work though. I vividly recall being high up in a corporate tower in Canary Wharf. There had already been some anxious exchanges with the security guard, who was unsure about giving me an access pass (whilst I became increasingly worried about being late). Then, when I finally found myself in the meeting room, looking out across the city skyline with Kylie, I got a sense that she didn't have very much time. She was doing her best to help, but she answered quickly. Indeed, she was powering through my questions at the start and I faced an unhelpful combination of me being flustered (which meant I couldn't easily slow the pace down because I was not yet calm myself) and her being busy (which meant she was keen to get the interview over so she could return to her document). Occasionally the lawyers teased me about my questions (painting me as a crazy academic with too much time on his hands to think about matters that were silly and inconsequential). That, as I will discuss, could be revealing too. Staging a kind of (anonymous!) gossip about what others in the study were doing (or saying) could help. That got them interested. It highlighted how I was actively analysing how their lives went on and it grounded our talk in the 'juicy detail' of what their peers were currently doing. Based on all that, what I now discuss are some of the interactions that proved particularly revealing, and that helped me to understand how their lives were structured (based on how they spoke of them). If, as I suggested, saying, doing and thinking are often interlinked, this discussion aims to show how that linkage worked in this context and what it revealed about their relationship with outdoor greenspaces.

Bothered by a Bee

Martin is considering how the summer had gone. It hadn't been particularly hot, with few spells of extended good weather. And that made it hard for him to identify things that had changed (since we last spoke) in terms of his relationship to outdoor spaces during the working week. The summer had struggled to insinuate itself into his consciousness and, like many others in this first study, he found it hard to speak of its impact. We are at his home in North London, looking down onto the city from the lounge of his top-floor flat. I was hoping that this might put him in a better position to reflect on his working life – surveying the city in which he spent a lot of his time in both a literal and metaphorical sense. But I was, so it seemed, to be disappointed. Perhaps that should have been unsurprising in view of how there were various onsite services at work (and the station for the underground train that took him there was less than a five-minute walk from his home). Either way, at the time he was clearly racking his brains to provide me with anecdotes that would fit the bill in terms of the kinds of issues and topics that he knew by now I liked to discuss.

Suddenly one sprung to mind. There had been a recent meeting that he had heard people speak about – a meeting that had happened outside. Apparently (or, at least, so the story went at work), the colleagues involved had taken the (as it turned out) rather rash decision to eschew the standard meeting room in which they usually congregated. The weather was deemed 'quite nice' and it had been a bad summer – so they felt they should take a chance on going outside! Notebooks were gathered up and they set off for the nearest patch of grass (reassured that, since it wasn't lunchtime, they were more likely to find some space). But that positivity didn't last long. Apparently, one of the lawyers involved was approached ('attacked' in her terms) by a bee. This was so unsettling that she couldn't concentrate. She went back early and the others decided they would probably not do this again.

Another anecdote was also about a group trip outdoors – though this time it had a happier ending. It began with an injunction that was not of the choosing of those involved. In a different firm, I was being told about how a series of interviews for a junior legal position had started off. And that was with an abrupt halt. A fire alarm had sounded and they needed to evacuate. But their time was also precious (and interviews can be hard to rearrange). What should they do? Sally told me how they had decided to relocate their meeting in the nearby park. And then, to their delight, it had actually gone quite well. Indeed, the location actually played positively into the interview. The senior partner was apparently so impressed by how the candidate had coped with the challenge that he 'couldn't stop talking about it' afterwards. The candidate, Sally thought, eventually got the position (and their ability to handle the outside was no small part of that in her view).

What can we take from these two starting anecdotes? The first thing is that these instances were remarkable, that they warranted comment – they were so unlike how life usually went on in these offices that they pushed their way into the collective conversation as the employees at the two firms involved processed these events together. What was also the case was that both of those who presented these accounts to me (and, indeed, also those with whom I discussed these instances when I brought them up in other interviews) were disdainful of the characteristics that these stories were apparently exposing. They knew they belonged to this professional group (whether they liked it or not), but they also preferred to position themselves in our conversations as different, as a little apart from it. Indeed, it was probably because of how these anecdotes served this function that they had circulated within their offices and been made available for comment. They allowed these workers to let off some steam about how a collective office practice had come to envelop them. They also allowed them to stand personally apart from it and to see themselves as a little different (Wasn't it ridiculous she was so bothered! Isn't it crazy they were so impressed!).

In Practice and in Principle

> Well I think it's like the National Trust. I have never been to any of these places but I think that somebody should be doing that, I think that we should be preserving these things.

I'm talking again with Sally in the park near her work about the way in which greenspaces were, and ideally should be, organised in that part of town. She immediately said that the idea of having them was 'lovely' (as many others immediately said when I introduced this topic) – they were a self-evidently good thing and so we should keep them! She was getting into the flow of her thinking in an abstract sense about why she thought these spaces allowed you to take yourself away from your pressures and anxieties. Indeed, it was almost as if that process was happening before my eyes as her position on these matters slowly became clearer, both to her and to me. But I was also interested in her lived experience as much as in how she liked to think of things – how things played out 'in practice', as much as how they ideally were 'in principle'. So I took a chance on interject-ing with a question about how things actually went for her (despite being a little nervous about disrupting her thinking and our conversational rapport). Her an-swer, provided above, was to link the greenspace immediately outside her office to buildings and landscapes (often far away) that were currently protected by a charity. So an area just 100 metres away from the office in which she worked was seemingly so disconnected from her experience that it might as well be in the same category of national resource as a stately home in rural England. In a sense, they were alike in being nothing to do with her (even though we were taking in a park at that very moment) – despite the availability of one and the distance to the other.

The same tension was evident in how other conversations went on (even with the same person) – how certain sequences of exchange between us took us in dif-ferent directions (how, in effect, different trains of thought carried them towards different understandings of their lives). Talking with each person more than once helped with this. If we had explored different ways of thinking about their every-day outdoor relations in a single interview some of this group would, I think, have become immediately aware of the inconsistencies in their 'opinion' as we jumped between different ways of looking at how things worked out for them (they were lawyers, after all, who were quite at home when generating and defending the 'views' they were professionally required to 'take' elsewhere). But, with a gap of three months between each conversation with the same person, I could explore different approaches (safe in the knowledge that they had often largely forgot-ten about the detail of how they presented themselves to me last time around). A good example came when I was talking with Emma, one of the most consid-ered and calm of this group (who had developed an impressively stoical stance on the long hours that she was labouring in her office). I suspected that taking such a stance helped her to handle the pressure – by repeatedly tacking back to the bigger picture, to what her job allowed her elsewhere in her life. In our first meeting in the lounge of her south London home, this approach was apparent. I was asking again about the idea of sitting out in the greenspaces outside her of-fice. Her response was that this was unlikely for her. But not impossible. She just didn't have the time. Then she stopped and added 'and they'd probably think of you as a fruitbat'.

Fast forward to our third meeting together and a rather different way of looking at this situation is on the table as we sat in a coffeehouse around the corner from her office. We'd been thinking about the issue of workplace dynamics – matters of how they should present themselves at work and ideas of being 'professional' at work (and how both of these could serve to push the idea of outside experience further away from their minds). I asked: 'what if a boss saw you on a bench alone outside?' That triggered quite a different response:

> I am not going to justify why I am sitting outside. I think that, if anything, people should be encouraged to go outside and eat their sandwiches and have a break. Because, you know, it clearly shows the ability to stop and to think … and it is in danger!

This was a rare outburst of emotion in our discussions together (and 'it is in danger' was quite an impassioned phrase to be using). Indeed, the force with which she expressed herself compelled me, in the moment, to agree with her. But it was also another example (like Sally's National Trust analogy) of the tension that they lived with – a tension between how they (often quite strongly, when they came to think of it) thought things 'should be' and the gradual analysis of the workplace dispositions they had quietly come to acquire.

At the end of all my interviews with the office workers, we all came together. This was both a chance for them to meet each other and an opportunity for me to put some final questions to the group as a whole. We assembled around a clear glass table in the dining room of one helpful respondent in East London. There was wine and crisps. One thing that I asked them to do to 'get the discussion going' was to rank a series of statements about what those interested in encouraging city workers such as themselves to have a fuller relationship with outdoor greenspace should consider. Low scores were frequently allocated to certain social dimensions. Certainly, the processes by which a collective mindset could gradually subsume them (as they were recruited into a shared practice of working) were downplayed. A number of other factors – about time and about providing places to go and sit outdoors – were deemed of much greater importance. But this, I think, also said something about the situation – how they, when there was some time and energy to reflect, thought of themselves. When you have the time and energy to reflect upon your life, you are a different person – you have a different relationship to the practices that you carry to the relationship that is there when you are in the midst of reproducing them (you become the 'in principle' version of yourself and the boring 'in practice' equivalent is briefly put aside).

I also think that this was partly about how they were reluctant to reveal too much about certain aspects in front of some unfamiliar peers. Alice had been keen to tell me during the comparatively confessional exchange of our earlier interviews that city lawyers were a particular 'breed' (and that this was likely to make me paint a particular kind of picture in my final report). There was more going on here than simply being an 'office worker'. And, true enough, in a good number of my discussions 'support staff' were observed to have a distinct relationship to the outdoors. By contrast to the lawyers with whom I spoke, support staff were noticed as being different because they had more time (and, importantly, also more mental capacity) to engage with the outdoor environment (by making the most of sunny days or being surprisingly responsive with their clothing choices to the changing weather). The lawyers had, by contrast, been, as she also put it, more fully 'absorbed' (by which I think she both meant absorbed into the building so that this was where they were naturally found and absorbed into the norms and practices that characterised this group).

Lawyers, for her, were typified by conformity and conservative approaches – they worked hard and they toed the line. And that could keep them indoors. This was probably true. But what was also important here, and what was likely to be important irrespective of the group under scrutiny, was the difference between processes of evaluation from outside and the lived experience of everyday routines from within. There was a difference here between how they liked to think of themselves (of how things should be 'in principle') and how things actually were 'in practice'. In this sense, the important question was not one of whether I had the right 'breed' on my hands (in understanding office work more generally) but more about how 'breeds' (of whatever kind) come about and sustain themselves. Either way, by paying attention to how they orientated themselves to my questions, I came to understand how they might have ideally liked things to be with their greenspace relations, but in practice didn't have much call to consider any alternatives (until I came along and presented them).

Hot under the Collar

Keeping cool under pressure; not getting hot under the collar: both of these commonplace phrases allude to how, in contemporary professional office work, aspects of bodily control are linked to personal capability. More specifically, human bodies that are too hot are a sign of failure, indicating that the contemporary businessperson is no longer appropriately on top of things. In many ways, it is also increasingly easy to keep potentially unruly bodies in check since, within the offices that these respondents occupied, a good amount of energy was devoted to helping them perform this kind of physical professionalism through the provision of 'appropriate' ambient conditions that helped them stay 'cool under pressure'.

This could, of course, play into their relationship with outdoor environments since those hoping to 'feel professional' might understandably stay inside the spaces that sustained these feelings. I was talking with Jenny about the idea of meeting friends outside at lunchtime. Rehearsing the 'practice' and 'principle' distinction, she thought doing so was a nice idea. But 'why was it just an idea?', I asked. Bringing it back to the actuality of her experience, she reflected on her imagined return to the office afterwards. At this point, an apparently 'dangerous' feature of the outdoor lunchtime presented itself. The perceived risk related to how, if she allowed herself to start enjoying the experience, she might soon relax. She might even start to sweat. Though it was hard to be specific, something, she thought, might happen to her. She would, in effect, she worried, start to feel less 'in control'. So, on balance, she wouldn't go outside. It would be too hard to resume a more purposeful feeling and to get back into the workplace groove afterwards. Better to keep going and not get distracted. She could always meet them later, if they wanted to do that in the evening.

When Remarkable Happens

Samantha has just ordered a Thai beef salad in a restaurant near her building and is rattling through my questions. She's still in her 'work mode', I think (fair enough since she has to get 'back upstairs' in under an hour). I'm asking about workspace sweat (lucky her!) and how they collectively felt about that in her office. Like quite a few others, she had examples at hand. These, as they often did, related to senior figures at work. They, so they mused, were either more used to a time when sweat was more common in the workplace or they just didn't care so much anymore – they were 'established', so perhaps they had fewer worries about being seen to get 'hot under the collar' in the office? She was getting up and running with her analysis and then, so it seemed, suddenly hit upon a memory of something remarkable:

> Well you see you get these partners and they have this way of leaning back on their chairs and putting their arms behind their heads and you can see it there – crusty.

She meant the armpits – caked with a mix of deodorant and sweat that had solidified together over time. She said, 'I think I've said this before'. I didn't think she had, not to me at least. But it did feel like an account that had previously been given. Perhaps at work. In this sense, she had hit on another remarkable moment – a remembered experience that had entered her consciousness and then stayed there as something worthy of remark.

Emma had something similar to report, though the shoe was now firmly on the other foot. We had been talking about how nice it was that some of her colleagues at work managed to vary their clothes seasonally. It was interesting to me that she talked about this in terms of 'managing' – in the context of how they spent long hours inside air-conditioned spaces and how the expectation was that they'd often stay there, varying clothing with the seasons required an effort (otherwise it might easily disappear). It was lovely, she smiled, to see her secretary – who was taken to have more mental energy available in her working day to do this kind of 'managing' – wear brighter clothes in the summer. That was lovely for her to see because it helped her recall that she too might do something different on the weekend – 'maybe go out in the garden with her kids or something like that'. But the story now was about how she had a similar experience as Samantha's senior colleague. She was working with a junior lawyer in her office, they were under pressure with a deal and the sun was streaming in. She noticed her female colleague looking anxiously down. Emma was sweating 'too much' and this was, if not a source of full condemnation, a source of some anxiety. She deflected the situation quite neatly by joking with her colleague about this (oh, ha, do I have sweaty pits?). But she also remembered this situation and doing so was testament to how they were taken to need to be cool under pressure, to have their bodies, and by implication their ability to respond to any client issues that might pop up, fully under control.

So, what is deemed remarkable can serve to police the possibilities for their practice since stepping out of line with the usual workplace disposition can risk condemnation. In that respect, these stories worked in a different way to those about the bothersome bee or the new recruit who got the job because he could handle being outside. Then, by thinking about the gossipy workplace stories that respondents could recall, I learnt something about how they (and their peers) enjoyed positioning themselves as apart from the workplace norm. Now, through a similar exercise, I saw how they could also serve to keep them inside it.

When You Actually Think about It

I'm talking about Sally's journey into work. She didn't think about the outdoor environment a lot during the working day. In fact, she was determinedly indifferent to its effects (and actually took a degree of pride in not letting it influence her actions). Though things might have been different when it comes to sweat, were it to rain as she travelled to a meeting, she wouldn't scurry about 'after an umbrella'. She would just have to arrive wet. I thought her journey in might be different – there was perhaps less time pressure at this point, more opportunity to connect? Not so. The idea of being 'outside' in the city was entirely alien for her. She paused and said, 'I'm afraid I just don't think about it – most of the time I'm just thinking about whether I can cross the road and read my book more or less at the same time – and most of the time I can!' The same was true of Samantha, who, never shy of seeing the humour in the banality of my questions, said that she just didn't think she was really 'outside' when she was in the city – she was just manoeuvring around people and carving out a route between 'the big grey buildings'. That's a bit 'weird isn't it', she added, before signalling the end of this passing reflection to me with a short laugh – 'hah'.

Both Sally and Samantha were too preoccupied to even think of themselves as 'outside'. And there were few environmental cues too – this was quite far from their idea of 'nature' and those things that drew their attention when they were outside were the more pressing challenges of getting where they needed to. But still they were too preoccupied – Sally with her book; Samantha with thoughts about the day ahead. And when they talked to me about this as 'weird', it was weird because the outdoors was, for them, largely irrelevant. Both were, in the context of being asked about it, coming to reflect on how they had come to embody a lifestyle that was, on reflection in the circumstance of being asked, a bit odd.

Others spoke of aspects of their lives that suddenly seemed 'strange' in the interview. These realisations were often occasioned by what they took to be rather esoteric lines of questioning (that dragged them out of their usual grooves of thinking and working). Particularly long periods of hard work or when they had a particular deal to complete, for example, could render some of these office workers almost entirely oblivious to the outdoor environment. The outdoors was something they imagined would have probably taken up some of their mental space, but during these periods it didn't. Sometimes, they recalled how they struggled to know what to say when the supposedly common English topic of 'the weather' presented itself. They had not noticed it. And so they had nothing to add (even though they suspected that changes had been happening). Tom spoke of going outside the office one late summer evening for a few minutes. He went to accompany a colleague who wanted to smoke after a stressful bout of work. Suddenly they both realised that it 'must have been' quite hot that day. They felt the lingering heat and they recalled how someone had said 'on the news or something' that the weather had been good. Tom, tellingly, didn't stick with his office job for long. His bemusement about this kind of existence made him determined to get out (tellingly before, he worried, 'he got too used to it'). By the end of the project he had left London (giving him more time for the outdoor sports that he loved).

Sally had been working hard towards her professional goal of securing partnership. And techniques for coping with weather were the least of her concerns. Yet, in passing, she did notice how her dry-cleaning bill was higher than usual as a result of a recent spell of rain (that had muddied the path she initially took on leaving home in the morning). In this way, the weather was only indirectly noted. And the same could sometimes be true when it got cold. One example related to a new winter coat that a senior colleague of Martin had bought towards the end of winter. He had been talking about this in the office. But he was not doing so to underline his style or seasonal adaptation skills. Rather this was about the lack of them. The comedy he saw in his situation was associated with how he bought the new coat because he thought he had lost his old one a long time ago. But what he had actually done was leave his coat on the back of his office door and then forgotten about it.

In other words, he had been working so hard inside that months had passed without him thinking of enlisting the help of his existing coat when venturing outside.

This was such a long period that he forgot about it. And he was, in the moment, proud of that. Either way, whether the result was quitting your job because you re-alised this was not the life for you, or whether it was taken to illustrate your capacity for hard work, the point was that these matters required an external impetus to be considered – either from a burst of particularly 'remarkable' weather or from a series of questions from me. Otherwise they largely just kept going. That is, however, un-less (as we now explore) they 'made an effort'.

Making an Effort

Yes, sometimes myself and a couple of work friends like to go for a wander at lunchtimes.

Sounds idyllic, I thought, as Stefan leant back in his chair and continued mulling over his workplace culture (with a glint in his eye that suggested he was amused by both my study and some of the habits that he'd seen around the office). Like Samantha, he seemed to enjoy this 'on the spot' analysis as he rose, in the mo-ment, to the challenge of providing a rationale for actions he otherwise 'just did'. Stefan wasn't describing to me the kind of somnambulant purposefulness that I was expecting, and which others had offered up with a smiling resignation (that was impressive in terms of how it signalled their diligence). He was positioning himself differently. So, I thought I should look into this. On further inspection, his 'wander' was revealed to have clear parameters. And it was also under threat (perhaps 'in danger'?). First, it involved advance emailing when the potential for a sequence of failed attempts meant it was always on the brink of disappearing as an option (they were busy, and no one wanted to seem too keen). The wander also involved two laps of the square down the road from his office (in a part of central London where there were quite a few green spaces scattered around). There was, so he reflected, a nicer square slightly further away. But getting there would involve crossing a busy road. And it eventually fell out of favour. Oth-ers were also known to undertake such laps around squares – commonly whilst catching up with friends who worked in other offices. But rarely was there any talk of going there and sitting alone. Somehow this was just too hard to do if you were up and running with your 'work head on'. Also, you didn't want to be thought of as the 'fruitbat' that Emma described earlier on (and you didn't want to risk the bodily unravelling, or perhaps more rightly, relaxing, that would have made it hard to get that head 'back on').

When these office workers did manage to effect such escapes, it was most com-monly described in terms of 'making an effort'. But this was 'making an effort' to do something that, from our discussions, they generally enjoyed. Why then was it often presented as so effortful? This was seemingly because it meant struggling free from the quicksand of otherwise increasingly habitual patterns of thinking and doing that were typical of their indoor office work, and which, as Alice tellingly observed, could easily 'absorb' you. And, in terms of how these 'escapes' were spoken of – both in terms of communicating it to others and legitimising the situation to themselves – having a kind of 'outdoor alibi' could help. Rarely were they going for a stroll – in this sense, as will be discussed shortly, Stefan was talking in quite an atypical register. They were getting the dry cleaning done, running an errand, picking something up that they needed for the weekend, popping out to 'grab' some lunch, or quickly catching up with a friend. The last one could be the riskiest in terms of unsettling how their working lives were otherwise talked and thought about. But, if necessary, this could be legitimised by claiming this was a friend with whom they might feasibly benefit professionally from meeting (someone who could provide a useful second opinion on an issue, for example). This, for some, seemed to make it acceptably purposeful.

And that was certainly how they spoke of these actions with their peers and their clients. If Stefan had missed a call from a client, he would say he was 'just away from his desk' as an attractively vague phrase that was also called upon by a number of others. He would never say that he 'needed to go out for a walk'. He would certainly not say he had gone out for a stroll. Though he didn't, initially at least in our discussion, quite know why this was, this was eventually understood by us as one of the reasons why he liked to speak of 'going for a wander'. There was a degree of irony here – the phrasing was, albeit in a less than fully conscious way, signalling how 'going for a wander' was a difficult thing to do when 'making an effort' was already a challenge. Otherwise, it was fairly easy to stay 'stuck at work' in the sense that breaking free required too great a degree of mental effort. And this disposition had its impacts, just as the work that required immediate attention did.

Leah was, elsewhere in our interviews (and rather like how Emma had been), at pains to emphasise how she would defend her 'right' to go for a walk at lunchtime. But she also went for such lunchtime walks partly because she already 'had it in her mind throughout the morning that she would'. This made me wonder how that squared with the idea of 'being purposeful' that was coming through elsewhere in the study. She thought that she liked to do it because more widely in life she was quite 'an indoors person' (she didn't care about being outside on her journey into work really, she got her exercise through fitness tapes at home, and, if she was honest, she enjoyed the pub much more than the park). She worried about the stigma of this way of being and the lunchtime stroll served to redress that balance. It reassured her. But the point was also that having an apparent reason to hand legitimised the action – it gave her a reason to make the effort. Lacking purpose was problematic when 'sometimes you need a reason to go into the park'. This was like how Stefan's carefully calibrated 'wander' served to mock the prevalence of the purposeful state. This was about reminding him (and others) that they were still humans and not the 'battery chickens' that Martin worried I was coming to see them as – that they still could benefit from (if they were quite careful about it) being less than entirely purposeful during the working day.

The Outside Coming In

Emma liked listening to the weather forecast. It was 'transporting', she said, to think about what was happening outside. But that rarely influenced what she did in her working life. Rather she responded to whatever the outdoor environment presented when she was called upon to step outside (one example that she particularly enjoyed recalling related to a time in her working life when she found she could take the top down on her car on her way home and smell the different foods that were being prepared in different parts of the city as she went). Stefan liked the weather forecast too. I asked why. He immediately said it was helpful as he presented himself as a resourceful individual who actively did things with the information that he acquired (and probably didn't bother to acquire it unless it was useful to him in some way). I took a chance and pushed him a little on why and how it was helpful. His response was that it might prove helpful because he might well be travelling across the country to a client meeting and he'd benefit from knowing what he needed to take. But in everyday life, he didn't need this information (and it must be said that the client trip that occurred to him as a justification happened a number of months before).

Whilst the idea of knowing (or, perhaps more rightly, thinking) about outdoor conditions was talked about as an enjoyable (somehow relaxing, somehow transporting) prospect, their everyday practice meant this did not happen very often. Part of the reason for this was of course (as already discussed) because the environmental conditions with which they were furnished at work had been so effectively engineered (depending on how you see it) to take them away from these kinds of 'distraction'. And that was regardless of whether you see them as enjoyable or unsettling. Indeed, there was very little to say about the layout, experience and design of their offices in a general sense – these were 'unremarkable' spaces (perhaps in spite of, perhaps because of, the amount of time they spent within them). They provided them with similarly beige surroundings, irrespective of the firm. They were used to them and they were busy. And so they didn't want to think much about them.

Windows were different. Or, at least, they were one of the features of their workplaces that were easier to reflect upon. Once you found yourself at a desk with a view, that could be hard to give up, said Stefan. As we sat in a meeting room whose widows faced directly onto a wall, I found it hard to disagree. There was also a status attached to having a view. They felt that views were beneficial and that is why they had this status. And these respondents definitely did 'find themselves' looking out of their windows. Those with desks facing inwards would also find themselves swivelling around on their chairs so they could do this too. Some, if they could, would even gravitate towards the window whilst they were making a client call – they'd stand up and drift towards the window (perhaps the view could make them think better – to see the 'bigger picture' in all senses). Though explanations were rarely at their fingertips, windows were a 'simple pleasure', something 'vaguely pleasant'.

Weather too (via these windows) sometimes managed to infiltrate their thoughts, and that was also a source of remark. Pleasure was definitely taken from the passing opportunity to be distracted by the elements (or, perhaps more rightly, by the way that the weather could still manage to puncture their working practice). But this was talked about in a particular way too – this was not about responding to any overdue 'nature reconnection' opportunity. Rather it was about a childlike sense of fun as this group (whose members, as we have seen, knew their working practices could draw them into the valuation of 'purposeful' states) momentarily allowed themselves to admit that there was 'more to life' than efficiently getting their jobs done at work. These things were 'bigger' than their immediate problems and they reminded them that they were part of a 'wider world' that put the achievement of professional ambitions into perspective. Alice took some delight in noticing how senior male colleagues suddenly become like 'big kids' if it started to snow or hail (or do something else dramatic with the weather) outside. But, even then, this rarely practically influenced them. In that sense, it was like listening to the forecast – enjoyable, but not often acted upon.

Other Ways to Escape

One benefit of talking to people on four repeated occasions (and talking to them alongside their peers as the process went along) related to how this made for a kind of gossip that both kept their interest (as discussed earlier) and provided me with effective topics (to write onto the sheet of paper described in Chapter 2). After I'd met all of the office workers once, and the second round of interviews was starting to loom on the horizon, I set about identifying promising lines of questioning (based on what they had already said). One of the questions that I posed was about whether they would rather a guaranteed lunch hour (in which they could go out into the greenspaces around their office or do something else) or whether they'd prefer a half-day off each week to do something similar wherever they liked. In doing so, I was starting to think about how, and when, the practice of office work might feasibly be tinkered with in ways that made for more outdoor benefits.

Everyone picked the second option. This was because stopping and having a restful break at lunchtime during the working week was often thought unachievable when, because of an amalgam of all the above issues, any kind of real escape from the workplace disposition was impossible. Unlike how it would be during a longer period in another place, during the week these lawyers had, in effect, been too fully captured by the practice of office work.

Conclusion

Overlooking Greenspace

The above excerpts were picked to give a sense of some of the most telling interactions that took place during the interviews involved in this first case study, along with an appreciation of how office work, as a practice, was living through some of these respondents. But what are the implications of these findings for how we think about fostering positive greenspace connections for what, to go back to the review that began this chapter, is a both a sizeable group and one which had not previously been studied in this way? Despite the various benefits that others have researched, and which these respondents recognised themselves when called upon to consider them, more often than not it was the case that greenspaces, and all the potential rewards of going into them, were pushed out of their consciousness in everyday life. There was little sense here of a residual evolutionary urge or deep-seated desire for spending time with plants and trees that was repeatedly thwarted by the demanding and needy practice of office work. In other words, they had adapted. And that was perhaps to be expected. But it is also worth thinking about how best to respond.

The younger respondents found this situation 'strange' more than those who had been working inside offices for longer. They were still evaluating what it was to do this kind of job, to allow themselves to become 'absorbed' into this profession. They had more thoughts immediately at hand within our discussions since they were still thinking about whether they could handle this, and how they'd do so in the future. The more senior respondents were often no longer thinking about such matters. They were just getting on with their 'work' in a way that takes us back to some of the discussions in Chapter 2 about how the use of particular terms discourages further inquiry or active consideration.

In this study, I was seldom involved in lengthy discussions about how the parks and gardens, often scattered around their offices, could be better designed to be more welcoming to workers like themselves. And that was partly deliberate. I am sure these respondents could have generated thoughts on these matters (and that redesigning these spaces in line with these thoughts could pique the interest of some of them and tempt them outside). But the point was rather to explore the extent to which they were already thinking about these topics. And the truth of the matter was that they seldom were. Their practice was effectively encouraging them to overlook local greenspaces. So, as I see it, one task for those hoping to encourage office workers to derive more restoration from their local greenspaces is about helping them to remember that these spaces are there in the first place. Or, perhaps more rightly, their presence was about remembering they could go, that their presence was something they might feasibly act upon when they were otherwise preoccupied.

As a result of talking to a range of interested parties about the implications of my office-worker conversations, there was a discussion in the local media about developing a 'text message service' (the study took place a while ago when smart-phones were rare even for them) that would do exactly that. The idea here was that this might usefully alert these workers to the times when particular plants were in flower or when events were happening in the greenspaces around their offices. The thinking was that these spaces could otherwise easily stay forgotten. Either way, and linking back to the Danish study in which their office workers said there was 'no real reason' why they did not go into greenspaces more during the working day, the point is that there actually was a reason, but it was one that was slippery and challenging to explore in a survey. That reason was, I think, the same as that which prompted their answer – they were preoccupied. And their preoccupations discouraged them from either going to these places or saying or thinking more about why they didn't. And so the overall way in which I would characterise the process that I saw in the first study was about 'forgetting the outdoors' – greenspaces were attractive in principle, but, in practice, the idea of going into them had been pushed out of the picture. And, by studying the subtle-ties of how the office-work practice gradually took over, I was able to reconsider the process of making everyday lives and greenspaces connect (or, at least, get a better sense of the challenges involved for those who would like them to).

Apparently Idle Talk from the Office

Beyond what could be practically done with the results of this first study, and going back to my second wager, I was also interested in what each study revealed about the relationship between talk and practices. More specifically, I wanted to consider how interviews focused on the everyday lives of each group illuminated these connections. Indeed, I had identified three suggestions that I wanted to question through my conversations with them:

- The first was that practices call forth particular speech patterns and, through these means, become more fixed and difficult to shake off. In some respects, I have learnt little about this here. But, in others, I have learnt something quite valuable. Certainly, I had limited access to the kinds of routine talk linked to this particular workplace practice. I had little chance to overhear much of the ongoing hum of interaction that partly comprised this form of work. I did, however, derive a useful sense of when the hum was disrupted. For example, when an incident becomes 'remarkable' that tells us something, I think, about actions and ideas that are alien to the practice at hand. So, by noting what becomes worthy of comment in particular social contexts (when an event becomes elevated into an anecdote for others in the office or for me in an interview) we can notice what kinds of event are not part of the standard talk that comes along with the practice. Having said that, I did occasionally get glimpses of how particular forms of talk played their part in sustaining this particular form of indoor working. The use of 'just stepping away from the desk' (as though the person should always be there) was one example. Another, though this worked in a slightly different way, was the idea of 'going for a wander' at work (through the situational recognition that this is out of step with how the office work practice gradually draws the person into a purposeful state). They are two sides of the same coin in that both indicate an orientation towards the practice at hand – one is what the practice encourages; the other a knowing nod to how certain actions are precisely not what the practice wants.

- The second was that how people handle the experience of being asked about the practice they reproduce is instructive (how willing are they to admit they are carried along by it, when do they decide to position themselves differently etc.). One of the most interesting features of the above conversations in this respect related to how exploring comparable topics could end up producing some very different responses (depending on how questions were posed and the line of thinking that the respondent was pursuing at the time). It can, for example, sometimes be hard to admit that you are not often thinking or actively making a decision about certain everyday actions (how to handle the weather, whether to go outside etc.) when you are, for interesting reasons, disinclined to see yourself as such. But there were also times when these office workers were entirely willing to acknowledge their lack of mental capacity for responding. For example, if someone says something is weird 'when you actually think about it', they are telling you that they don't often think about it (and they are willingly doing so). This said something about the benefits of slowly circling the practice in order to fully understand how it is handled (both in interviews with me and in people's own ways of managing everyday life).
- The third was that people often reach for the suggestion of a social norm under the circumstances of being asked their practices. This was evident here. But this was not a case of this being a strategy for legitimising the actions of the individual at hand (I'm doing the same as everyone else, so leave me alone!). Rather what was interesting here was how some of these workers positioned themselves as apart – making observations on others as much as admitting to being equally subservient to the practice themselves. Though they spoke of a 'breed' of people in their offices, this was not one to which they eagerly admitted belonging. So, people can observe group norms, and they can act in line with them, without entirely seeing themselves as belonging to them.

Furthermore, they can also get excited about times when such norms are disrupted. This was why, when some of their peers respond like a 'big kid' to outdoor weather, that was welcome: it pointed to alternative ways of relating to the outdoors. The collective did certain things, but there were still exceptions. And these exceptions were, in this case by and large, to be celebrated (partly because it was often quietly acknowledged that doing something different was difficult).

Lessons have now been learnt about both how outdoor experiences are handled by professional office workers and what their ways of speaking about work reveals. But, in other domains of everyday life, things are very different. Office work, after all, was always going to be a tricky practice to unsettle when it is, by definition, attached to an indoor environment. Furthermore, and to state the obvious, this is also 'work'. People are paid to submit to this practice so there is good financial reason to fall into line with the demands that it makes. Other practices meanwhile are much more mobile and, if we turn to activities that are traditionally thought of as leisure, there will surely be less of a sense of stricture, of feeling that we should give ourselves up to them? My next chapter explores whether that is the case by turning to another popular, and increasingly widespread, practice in the United Kingdom and many other counties, namely recreational running. My next set of questions are about how the experience of running draws people into particular running environments, both indoors and out, and what can be learnt from speaking with some of those who regularly run within them.

References

Baldry, C., Bain, P., and Taylor, P. (1998). 'Bright Satanic Offices': intensification, control and team Taylorism. In: *Workplaces of the Future. Critical Perspectives on Work and Organisations* (ed. P. Thompson and C. Warhurst), 163–183. London: Palgrave.

Bringslimark, T., Hartig, T., and Gridal Patil, G. (2011). Adaptation to windowlessness: do office workers compensate for a lack of visual access to the outdoors? *Environment and Behavior* 43: 469–487.

Chang, C.-Y. and Chen, P.-K. (2005). Human response to window views and indoor plants in the workplace. *HortScience* 40 (5): 1354–1359.

City of London Corporation. (2013). *Green Spaces: The Benefits for London*. London: City of London Corporation.

Colley, K., Brown, C., and Montarzino, A. (2016). Restorative wildscapes at work: an investigation of the wellbeing benefits of greenspace at urban fringe business sites using 'go-along' interviews. *Landscape Research* 41: 598–615.

de Decker, K. (2016). The curse of the modern office. Low Tech Magazine. https://www.lowtechmagazine.com/2016/11/the-curse-of-the-modern-office.html (accessed 27 January 2021).

Gilchrist, K., Brown, C., and Montarzino, A. (2015). Workplace settings and wellbeing: greenspace use and views contribute to employee wellbeing at peri-urban business sites. *Landscape and Urban Planning* 38: 32–40.

Gladwell, V. and Brown, D. (2016). Green exercise in the workplace. In: *Green Exercise: Linking Nature, Health and Wellbeing* (ed. J. Barton, R. Bragg, C. Wood et al.), 139–149. London: Routledge.

Hampton, S. (2017). An ethnography of energy demand and working from home: exploring the affective dimensions of social practice in the United Kingdom. *Energy Research and Social Science* 28: 1–10.

Hartig, T. (2006). Where best to take a booster break? *American Journal of Preventative Medicine* 31: 4.

Heerwagen, J. and Orians, G. (1986). Adaptation to windowlessness: a study of the use of visual décor in windowed and windowless offices. *Environment and Behavior* 18: 623–639.

Ingold, T. (2004). Culture on the ground: the world perceived through the feet. *Journal of Material Culture* 9: 315–340.

Kaplan, R. (1993). The role of nature in the context of the workplace. *Landscape and Urban Planning* 26: 193–201.

Kaplan, R. (2007). Employees' reactions to nearby nature at their workplace: the wild and the tame. *Landscape and Urban Planning* 82: 17–24.

Kweon, B., Ulrich, R., Walker, V. et al. (2008). Anger and stress: the role of landscape posters in an office setting. *Environment and Behavior* 40: 355–381.

Larsen, L., Adams, J., Deal, B. et al. (1998). Plants in the workplace: the effects of plant density on productivity, attitudes, and perceptions. *Environment and Behavior* 30: 261–281.

Lee, S. and Brand, J. (2005). Effects of control over office workspace on perceptions of the work environment and work outcomes. *Journal of Environmental Psychology* 25: 323–333.

Lottrup, L., Grahn, P., and Stigsdotter, U. (2013). Workplace greenery and perceived level of stress: benefits of access to a green outdoor environment at the workplace. *Landscape and Urban Planning* 110: 5–11.

Lottrup, L., Stigsdotter, U., Meilby, H. et al. (2012). Associations between use, activities and characteristics of the outdoor environment at workplaces. *Urban Forestry and Urban Greening* 11: 159–168.

Murphy, M. (2006). *Sick Building Syndrome and the Problem of Uncertainty*. Durham, NC: Duke University Press.

Shibata, S. and Suzuki, N. (2002). Effects of the foliage plant on task performance and mood. *Journal of Environmental Psychology* 8: 265–272.

Shove, E. (2003). *Comfort, Cleanliness and Convenience: The Social Organisation of Normality*. Oxford: Berg.

Strom, S. (1994). *Beyond the Typewriter: Gender, Class, and the Origins of Modern American Office Work, 1900–1930*. Champaign, IL: University of Illinois Press.

Thomsen, J., Sønderstrup-Andersen, H., and Muller, R. (2011). People–plant relationships in an office workplace: perceived benefits for the workplace and employees. *HortScience* 46: 744–752.

Van-Meel, J. (2000). *The European Office: Office Design and National Context*. Rotterdam: 010 Publishers.

Chapter Four
Avoiding the Outdoors: On the Treadmill

Running Away from the Benefits of Running

Around 2.5 million years ago, the forebears of the modern human started to venture away from the safety and sustenance provided by trees. In doing so, they set in train a series of evolutionary changes that ended up making them peculiarly good at a kind of 'persistence hunting' that involves tracking animals until the animals became exhausted and vulnerable to attack (Carrier 1984; Liebenberg 2006). These included standing up on two feet, the arrival of more efficient sweating methods that were also linked to the slow disappearance of fur, and an increasingly mobile neck and head arrangement that made for an unwavering visual focus as they chased after this new set of moving targets (Raichlen, Armstrong, and Lieberman 2011; Mattson 2012). Some evolutionary anthropologists have even gone as far as to suggest that gaining access to animals, as a hitherto untapped source of protein, served to fuel the development of the peculiarly large, energy-hungry brains that modern humans now carry. In effect, their idea is that our ancestors ran towards the intelligence levels required to survive in the comparatively complex societies of today (Lieberman 2013).

In light of this context, it is reasonable to suggest that running should still be relatively instinctive for people. We have always done it (indeed, it may have made us who we are), and, given the slow creep of evolutionary change, we might assume that modern humans are still partly hardwired for running

The Unsettling Outdoors: Environmental Estrangement in Everyday Life, First Edition. Russell Hitchings.
© 2021 Royal Geographical Society (with the Institute of British Geographers). Published 2021 by John Wiley & Sons Ltd.

(McDougall 2010). So, linking back to Chapter 3, whilst office work has become so ubiquitous only over recent decades, our relationship with running goes back much further. Yet, having said that, it has also been observed how recent changes connected to the idea of providing people with 'comforts' akin to those of the indoor air-conditioned office could mean that lifestyles are often increasingly misaligned with this legacy. In other words, many of those in today's societies no longer have the opportunity or (perhaps more correctly) the inclination to do the running that may have served as our evolutionary engine. And, though we cannot reasonably expect members of modern societies to run regularly for hours like their 'persistence hunter' ancestors, evidence is still mounting to suggest that the lifestyles that are coming to predominate (in wealthy, urban, western societies, at least) may have drifted too far from the conditions to which we are still physiologically adapted (Mattson 2012). This, at least, is the basic contention of the 'mismatch theory' of human development (Malina and Little 2008) – a term that highlights the apparent irony of how our evolutionary success has left us with lifestyles that superficially seem easier, but to which we are actually ill suited.

For some in this field, even when people run, this now often happens in less than ideal ways. One aspect relates to how commercial interests and another set of beliefs about providing comfort have gradually encased running feet in specialist shoes that promise performance but may actually impede the establishment of effective technique (Jungers 2010; Tam et al. 2013). Another relates to how modern runners might do well to remember that endurance running was originally associated with sporadic hunting bursts. We are accordingly not well suited to running too often (O'Keefe et al. 2010; Boullosa et al. 2013) as that leaves insufficient recovery time. In that sense, running is once again akin to office work in the way that, as a practice, we might benefit from considering how routine kicks in, and how attempts at engineering the ideal experience can have unforeseen consequences.

That takes us nicely to the focus for my second case study, which is concerned with the 'where' of contemporary recreational running. As I will discuss, and in a way that should come as no surprise in view of what has already been said, greater health benefits seem to be accrued if running happens in certain kinds of outdoor environment. Again, this may link back to how our ancestors were seemingly not only designed for running, but also for running through and across certain landscapes (O'Keefe et al. 2011). For this chapter, this raises the question of why it is that some of us have ended up on indoor treadmills whilst others remain in apparently more 'natural' environments outside. In order to start to answer it, I will begin with what market research has suggested about how and where people run today. Britons, for example, have been running more than ever. But this was not because of any residual evolutionary urge. It was more about how busy they had become.

The Changing Places of Recreational Running

It is wrong to infer a societal interest in exercise from sports kit sales figures. This is because wearing these items is often more about communicating style and sub-culture than actual involvement in sport. In other words, many people buy trainers and tracksuits because they think they look good, rather than because they help them to exercise more effectively. Nonetheless, when people buy exercise footwear in the United Kingdom, the activity to which they now most commonly refer would seem to be running (MINTEL 2013). We might conclude that running is now the default exercise option in this country. Certainly, there is evidence to suggest that recreational running is happening more now here than ever before as, in contradiction to some of the anxieties of those who were worried about the evolutionary mismatch thesis, between 1999 and 2009, for example, an additional 10% of the population expressed an interest in running (MINTEL 2010). Many Britons seem to have acted on this too. In 2012, the most popular fitness activity was running since, whilst many shied away from the title of 'running' (preferring, for reasons to which I'll return, the reduced pressure of 'jogging'), almost 9% of Britons claimed to run regularly (Keynote 2013). And interest in this form of 'keep fit sport' has shown little sign of dwindling since (MINTEL 2018). As such, and as England Athletics (2013) has been keen to point out in the past, running could now be considered an obvious starting point for those who are hoping to foster public health through regular exercise in the United Kingdom. In response to widespread worries about sedentary societies, they argue that, if Britons are already running more and more, the task is merely one of amplifying an existing social trend.

In terms of outdoor running, a number of explanations have been offered for this. One of the most popular relates to the decline of collective schedules and the amounts of free time that people feel they have (Henley Centre 2005; ONS 2011). Unlike running, many other exercise options require co-ordination to be sure that participants can take part together. And many Britons now seem doubtful about being able to co-ordinate. As such, the comparative beauty of running is that it can be more easily inserted into the schedules of those who are no longer willing to commit to collective sporting activities (MINTEL 2010, 2013). In this respect, running further benefits from being what some have called a 'doorstep' (MINTEL 2010) or 'unstructured' (England Athletics 2013) exercise option in the sense that it does not require travelling to a dedicated environment and can instead begin as soon as we step outside. Connected to this suggestion, a further broader shift has also been discerned from 'sport' to 'exercise' (Keynote 2010; MINTEL 2010) as the central focus of exercising Britons has evolved from competition and camara-derie to individual fitness and performance. Finally, there are more prosaic mat-ters to do with how, compared to sport (which often requires physical equipment and bookable spaces), outdoor running is cheap (MINTEL 2010) – especially so, if, to return to my first observation about recent trends in sports kit sales figures, many Britons now already have the right shoes to hand.

Data on indoor running – generally involving treadmills, though gyms occasionally boast indoor tracks – is harder to find. This is partly because doing so is often combined with other activities and partly because of an intriguing reluctance (to which I will return) to classify indoor running as 'running' (Keynote 2013). Nonetheless, there has clearly been a massive growth in UK gym attendance (Keynote 2013; IBIS 2018) as these facilities have morphed from the preserve of affluent elites and determined bodybuilders to a widespread means of working towards health and attractiveness (Keynote 2010). In 2013, the United Kingdom was second only to the United States in terms of likely gym membership (Keynote 2013) with this almost doubling between 1998 and 2002. Though, as with the above story of sports kit sales, membership does not always mean activity, in 2016, about 15% of Britons had gym membership. And the market is predicted to grow (IBIS 2018). In terms of what people do there, cardiovascular machines, such as the treadmill, would seem to be an important part of the experience. In 2012, 6.6% of people in the United Kingdom said they used these machines either 'occasionally' or 'on a regular basis' (Keynote 2013). Some have even put the growth of the personal fitness equipment market down to how those now familiar with technologies like the gym treadmill have started to see the appeal of something similar at home (in 2009, 18% of UK adults owned such equipment according to Keynote (2010). In explaining these trends, many of the same accounts seem to apply indoors as out: people no longer feel they can commit to group activities and prefer to find the most personally convenient ways of achieving individual 'fitness' rather than attempting to take part in collective sport (Keynote 2010). However, another intriguing suggestion is that indoor gyms are attractive partly because they 'eliminate the problem of unpredictable weather' (IBIS 2019).

There are indications that running is increasingly becoming the obvious exercise choice in the United Kingdom. Yet, though Britons seem to be running a lot (and running in various environments), less is known about the processes by which they become attached to some instead of others. Compared to office work, running is, to state an obvious but important point, more mobile. We also know that people are currently exercising in countries like the United Kingdom in a diversity of environments. According to one survey, in 2010 there was a roughly even split of Britons exercising in the sports centre (27%), the local park (32%) and the gym (27%) (MINTEL 2010). Though this can be interpreted in different ways, one way would be to suggest the United Kingdom was then at a crossroads in terms of where exercise most often happens (I was unable to find more recent statistics so cannot say which path it has taken since).

It is, however, quite possible to picture the future migration of running indoors as the idea of exercising in dedicated environments that offer a reassuringly predictable experience takes hold. Certainly, it is never a done deal that identified exercise forms will stay where they currently are. Tennis, for example, initially took place indoors before eventually moving outside (Eichberg 1986) and swimming increasingly happens in indoor pools in the United Kingdom (Ward 2017). What we can be sure of is that, if our exercise practices relocate, that will happen gradually. And such relocations will be the result of an accumulation of actions that seem fairly inconsequential at the time. This raises the question of how running ends up happening indoors rather than outside (along with, to return to my particular interest here, the implications of any relocations for potential greenspace benefits). But, before we get into that, and as with my other case studies, it is worth thinking about existing work on the greenspace benefits of exercise in order to establish how a focus on these processes might enrich this work.

Green Exercise and Lived Experience

The most common means of evaluating the comparative benefits of indoor and outdoor exercise has focused on the effects of outdoor activity in the presence of vegetation as a phenomenon that has since been dubbed 'green exercise' (Pretty et al. 2005; Mackay and Neill 2010; Gladwell et al. 2013). These valuable studies sometimes draw on the previously explained evolutionary framework to test the hypothesis that exercise may be better for us if it happens in the company of plants and trees. One argument, for example, relates to how, since early humans inhabited grassland environments, their modern descendants might retain an affinity for exercise within vegetated spaces (Gladwell et al. 2013). And there is now another sizeable evidence base to support this. For example, either looking at or being in green environments when running can seemingly augment the more general health benefits of exercise by improving mood and self-esteem (for example, Pretty et al. 2007; Akers et al. 2012) and lowering blood pressure (for example, Pretty et al. 2005; Park et al. 2010). There may even be further benefits if people end up running for longer in greenspace because the meditative effect can distract from the monotony (Gladwell et al. 2013). In short, whilst running is good for us, running outside around greenery seems even better.

These studies commonly involve the manipulation of identified features of the experience whilst others are held constant. One strategy, for example, has been to put people on indoor treadmills or exercise bikes and then vary the images they see before them as they exercise in order, for example, to compare the effects of looking at city streets or green parks (Pretty et al. 2005; Akers et al. 2012). If we buy into the suggestion of a standard human response (on this see Thompson Coon et al. 2011), these studies make a powerful case for furnishing city dwellers with environments that help them to benefit from 'green exercise'. However, what others do with this kind of evidence is another matter. Such studies can, for example, support an argument for creating virtual natural environments indoors

(Depledge, Stone, and Bird 2011). If ageing societies, for example, find real-world equivalents difficult to negotiate (grass can get slippery, older people might trip over roots or irregular paths), such innovations might make a lot of sense in terms of fostering public health in the future. But such scenarios also hint at the irony of how studies designed with a view to encouraging outdoor activity could end up having the opposite effect if they are used to justify the replication of indoor experiences that were originally merely part of the research design.

Working through such scenarios encourages us to consider the wider cultural processes that stand to shape the social future of green exercise (Bamberg, Hitchings, and Latham 2018). We have seen glimpses of the pleasures associated with indoor exercise such that, though the outdoors may be 'energising', the indoors can, for some, feel more relaxing (Plante et al. 2006). Then there are social differences such that some of those on lower incomes, for example, find indoor exercise especially appealing despite being less able to afford it (Burton, Khan, and Brown 2012). In any case, it may be worth thinking less often about the ideal environments for beneficial exercise outdoors, and more about how different sets of people and particular exercising places become gradually entwined in their everyday lives. In the interview research of Howe and Morris (2008, p. 319), for example, running through woods was valued by their respondents because of the 'rolling dirt track ... that is soft underfoot, being occasionally covered with bark chips in places and wooden planks in others'. Yet others can also recoil from these features. 'Natural' environments, with all their distracting variation, can, for example, be particularly unappealing for those with an unwavering focus on achieving optimum athletic performance when they exercise (Butryn and Masucci 2009).

Speaking with Some 'Non-runner Runners' in London

This chapter now turns to a study involving two sets of 10 runners in London. Both ran for at least 20 minutes at least three times per week. Though some took part in races, both groups were happy to be called 'recreational runners'. In these senses, they had much in common. The difference was that one half generally did their running outside and the other half was generally found on indoor treadmills. I should emphasise at this point that the outdoor runners didn't always enjoy a wonderfully untroubled experience of gliding through local parks when they ran. Many of them had no choice but to run along streets that required them to negotiate traffic, people and pollution. But all of the outdoor runners managed to do at least some of their running in, or at least alongside, greenspace.

In recruitment terms, I might have initially thought the task would be an easy one – given that many people were running, given that they presumably decided at some point to take it up (unlike being in an office, which is more a means of making money), and given that they might feasibly feel quite proud of their running because of its health benefits and associations, I might have expected to easily attract enthusiastic volunteers. However, in actuality, there were recruitment challenges. These related to both how recreational runners saw themselves and how they preferred to speak about their running.

First, potential recruits often sidestepped the suggestion of having the right attributes for a 'running study'. This reaction is probably a feature of recruitment for interview projects more generally. Though it is seldom discussed, potential respondents can, in this way, seize on arguments that allow them to escape the task of taking part whilst also painting themselves as the kind of person who'd definitely otherwise help. However, in this case, it was also because they didn't see themselves as, or (perhaps more rightly) hadn't yet needed to think of themselves as, 'proper runners'. Though many of those with whom we eventually spoke were frequently found pounding out their runs in London, they didn't initially think that was interesting (either to us or to them) – surely, we really wanted to talk with those who did competitions, with those who had done a marathon perhaps? In making such claims, they were hinting that other kinds of runners were 'the real thing'. Our answer, however, was that this was precisely why we wanted to spend time with them. It seemed, after all, that the recent resurgence of running in the United Kingdom was little to do with getting into running clubs or taking part in a more committed kind of competition. Rather, we figured, the resurgence was more about people 'just running' regularly. And so we went after the 'everywoman and everyman' of recreational running (Hitchings and Latham 2017).

Second, there was an interesting related issue to do with the extent to which such people were reticent about talking with us (or anyone) about their running. And that was pertinent to the themes of this book in view of the suggestion that the extent to which people are interested to talk about certain features of their practices is, rather than to be used as an argument about how foolhardy talking with them is, telling us something about how forms of everyday life are held in place. These participants were keen runners for sure (though perhaps 'keen' rather overplays their intention). However, for reasons that I will explore in this chapter, whether they were 'keen talkers' about their running was something else. Nonetheless, we eventually persuaded 20 people to help in our study and, after expressing some degree of reticence, all of them committed to the activity. Some (as we will see) even became interested in what their responses told them about themselves. In line with the current profile of those who run in the United Kingdom, most came from the middle classes, with our runners having jobs that ranged from research, to marketing, to policy and (again) to the law. For the outdoor runners, we also specified that we wanted to speak with those who could afford a gym membership and had a gym fairly near to hand (either close to their home or their work). So these were, according to popular ideas about what shapes our practices, those who could easily relocate their running, if they wanted to.

The next step was to get to know them (and their running) a little better. And, since we were interested in the relationship between what they did and how they spoke of it, we figured that it would make sense to start by doing some running with them. We were also well aware that (particularly since we were doing a comparative study) not all running is the same, even if some of those involved (as we will see) have good reason to present it as such. With that in mind, each individual was asked to start by letting us know when they were going for a run that was relatively 'typical' for them. Then one of us turned up, hopeful that we would be able to keep up (or, more often, that we'd eventually gain access to the relevant gym if they were running indoors!). The idea in doing so was to watch what they did, talk a little about how they did it, and then use this to do more effective interviews afterwards.

I must admit to some nervous handshakes in reception areas and fixed smiles at park entrances as we began 'taking an interest' in sequences of actions that were commonly undertaken either alone or with those who they knew much better. Still, we persevered and, as a result, ran alongside some 'non-runner runners' in a range of ways and places (as we looked across from adjacent treadmills or circled around the city). We spoke a little as we went, hoping that this would take us closer to the preoccupations they had at the time (the thoughts that started to run through their heads as their running began and then continued thereafter). Crucially, doing so also helped provide a sense of when, and how much, each respondent wanted to think and talk about certain features of their running. In an interview, after all, the expectation is that the conversation will go on. Albeit with interesting pauses and digressions, the idea is definitely that talk should ideally 'flow' until the end. When running with someone, by contrast, we were better placed to see when they wanted to stop the talking because they were (with various effects) being absorbed into the task or activity at hand. Furthermore, in this case, there were definitely also issues linked to the points at which exertion can make it hard to speak (as much as anything more academic to do with consciousness)! So, we talked a little with them at the start, and then ran for a relatively silent period with them. Then we spoke again after the run had ended about how it 'went' and the different actions and amendments that we thought we had observed.

A week or so later, the same pair met again and, drawing on what they had been seen to do and say in the first run, along with what the others had done and said in theirs, explored a variety of ways of characterising their running, considered different ways of thinking about running, and wondered about how their running might change as they moved into the future. It was nice to meet them again after having suddenly parachuted in their running lives at the first meeting – as though being granted access to an otherwise personal, and often solitary, experience had somehow bonded us. Despite that, and as will now be described, these were not always the easiest topics to handle in an interview exchange. But, in line with the arguments presented earlier, these difficulties told us something about how they related to their running. And so, drawing on how both groups spoke, this chapter now considers what this study revealed about the potential pull of the indoor treadmill.

The Three-mile Threshold

I'm running with Charlie, a medical sales trainee, on adjacent treadmills at her workplace gym. The pace is fixed, she is faced determinedly forward, and her focus seems unwavering. Her eyes appear to be trained on a point on the wall that (as revealed in our subsequent discussion) she 'just liked' staring at. Then she turns to me and says that she'll now 'see how she feels'. We are at mile three of the run. Buttons are pressed and speeds change – first faster, then slower, in a seemingly erratic way. And then we stop. She gulps from the bottle she had put in the treadmill holster, looks across, and smiles. It's done. We have been running for 45 minutes.

How is it that Charlie came to adopt this running pattern? After the accompanied run, I asked why her pace had changed at that point: what was she thinking? Her initial answer was 'not much' (in a way that already told me something about being careful not to assume an underpinning 'reason' for actions). Rather, the truth of the matter was that three miles had somehow, gradually over her recreational running career, become a milestone target that she compelled herself to reach. This was just what she did. After some reflection, as I nervously left the topic hanging in the air, she eventually unearthed a rationale: the three-mile threshold was born of how, when she first starting running after quitting smoking some years previously, she ran three miles outside before suddenly being sick. Since then, she always ran for at least that distance. Not getting there was not an option.

Charlie's threshold points to the serendipitous nature of how running routines, irrespective of environment, became entrenched in our study. Various objectives and approaches that initially arrived more or less by happenstance slowly became more fixed, more unshakeable. When asked about how they 'came to run' in the ways that they did, the most common response was one of relative bemusement. No-one was really thinking like that. Running was positioned to us as a straightforward activity whose basic form was so obvious that it made for a strange discussion topic. In one important sense (and going briefly back to the evolutionary perspective presented above), they were indicating that running was indeed something they naturally 'just did'. When asked about 'learning' to run, school experiences were most often reached for. This was, I think, partly because that was a time of learning in their lives and partly because this was when they were first exposed to the idea of 'exercise running'. They'd certainly done it before, but not with any real connection between doing so and the health and weight loss benefits that were now readily attached to the activity.

What was also clear was how this was about finding an answer in the exigency of the moment. Respondents had often been running for several years. And many spent significant amounts of their free time running. Yet, despite this clear level of commitment, for all but a minority, practical matters of technique went hitherto relatively untouched in terms of critical evaluation. Though runners of long-standing, the only point at which many had previously reflected on these details was in response to sudden disruptions: the arrival of injuries or noticing that their trainers had worn through (or under the conditions of being confronted by a strangely curious researcher). And this was often part of the attraction.

They Must Be Bad Because I Don't Use Them

Looking back over their running accounts, many respondents indicated that a common running objective was akin to setting up a running 'system' that 'worked for them' in the context of their wider lives at the time (and then submitting themselves to it). We saw this in Charlie's unwillingness to think about what she was doing until mile three. Another example came when talking with Rachel, one of the outdoor runners, about the idea of treadmills. She was adamant about her personal aversion to them. On following up on her reasoning for this, her response was telling: 'They must be bad because I don't use them.' The presumption here was that, at some point in the past, she must have made a good decision about that (though she had evidently forgotten about that process now). The interesting point here, however, was that she was not interested in exploring whether this intuition was true. Rachel, and others, had decided, in effect, that their current running practice 'knows best' – it had their best interests at heart (wherever it came from) and that was comforting. It was therefore better not to subject it to much, if any, critical scrutiny.

Questioning certain aspects of their running was explicitly not what they wanted to do – a somewhat surprising finding in view of how, as (eventual!) volunteers for an interview study, they were presumably more amenable to the idea of talking about their running than most. Yet this is not to say that those who took part in this second study lacked immediate running goals. Rather it is to emphasise how these goals remained within the parameters of their existing running practice – running a little further, going a little faster. It was not about troubling things in any more fundamental way. We tabled possible amendments with them through various means (Do you think you could run in this way? What would it be like to run in that environment?). But significant change was often off the cards because these runners didn't want to unsettle what they took to be a positive feature of their lives. Here we noted the occasional sigh of exhaustion at the end of our interviews. Samantha said she never wanted to think (or talk) about her running ever again (a difficult comment to handle in the moment when attempting to end the interview on a positive note)! And so, though these two groups of respondents had volunteered because they were, at least partly, intrigued by the idea of discussing their running, in actuality, many found it surprisingly challenging to evaluate certain aspects. Or, perhaps more accurately, there were certain procedural aspects they preferred to leave alone. And, in this respect, this was more intensely felt than for the office worker, even though we might imagine it to be more enjoyable to run than to work (and so more of an enticing prospect to speak of running instead of office life).

Someone Told Me Once

In a suburban gym, overlooking a car park with a fast-food restaurant and a homeware store on the other side, I'm talking with James about how he organised his treadmill runs. He's been doing them 'on-and-off' for a number of years now and his 'system' was working for him. Indeed, certain elements were the source of some pride. Well aware, for example, that he might easily become inclined to skip a session when it came to the crunch (especially after a long day at work), he always made sure he packed his kit on selected mornings. Then he had 'no choice' but to run after work. Otherwise he'd feel 'pretty silly' about bringing this stuff in and then eventually carrying it back home unused (extra exercise perhaps, but not of the intensity that he was after). We were talking about the 'incline' feature of the treadmill. As a widespread option, this gave him scope for tinkering with the settings or picking predefined 'types' of run (simulated mountain crossings, interval training with repeated bursts of speed). One of them was about changing the incline of the treadmill so that it became steeper, partly we can presume to simulate running up a hill (so there was evidence here of attempting to emulate the experience of being outside, with all its subtle and challenging undulations). None of the other treadmill users at his gym took advantage of this feature, he thought. But he did. Why? Because he thought this was better? That outdoor running was the ideal? No. It was 'because someone had told him once' that this was a good idea and so he went with that (and had been going with that ever since).

And that was similar to how active thought about improving their running experience or optimising their performance was again generally unwelcome in this study. Despite running for a number of years, and doing so fairly regularly, a standard feature of James's running practice was the product of a half-remembered piece of advice from years ago. This was a little like Charlie's water bottle in the holster (mentioned earlier), which was religiously filled up before she started her treadmill run but was thereafter often untouched during her actual run. Possibly someone had 'told her once' about the importance of having a water bottle to hand such that this had become part of her routine practice (even though the fact that it was untouched rendered it redundant). Simon always had an energy drink after his run because 'this seems like the kind of thing that you should have'. We asked everyone about their running shoes. Why had they picked those they currently used? Few answers were readily available. Despite the attempts of sports kit companies who may be very keen to tell us that we should really think about how our running is supported and cushioned by our trainers, these runners were rather uninterested in refining their practice in these ways. They were choices available to them, for sure, but the point was not to think about these things. The point was simply to keep going with what they were already doing.

Indeed, talk about their 'running kit' often came with some self-deprecation. When half-remembered statements about the promise of particular 'foams' or 'gels' in different parts of their chosen running shoes (or rather those shoes that they had ended up running in) were relayed to us, this could be done with a conspiratorial look. This was as if to suggest that we both probably knew that these features and technologies might not actually do all that much, but that everyone bought them anyway. This was also done because they couldn't always really remember the decision-making process, in any case. It was better to make an implicit appeal to a common experience of having to negotiate the half-truths offered up by manufacturers than to say they had 'no idea' (though Maddy did!) about why they were running in the (often fairly expensive) running shoes that they currently wore. That 'someone' who 'told them once' about the benefits was now the salesperson in the running shoe (or general sports) store. Otherwise they just kept going. On asking Mary about her running shoes, she looked puzzled. Then her face suddenly lit up – yes, she did have something to say about this! She exclaimed 'Oh my toe has started hurting recently.' She wasn't really answering the question. But she was trying her best to dig up thoughts and experiences that took her closer to an answer. And, though she hadn't yet given it much thought herself, remembering that her toe had started to hurt was the best she could do.

It's Just My Preference!

In this regard, responses were often along the lines of 'this is working for me' or 'this suits me for the moment' in what I'd argue was an improvised attempt to head off certain lines of further enquiry. On one level, the above could be read as slightly embarrassing for all the runners involved: they were doing things that they couldn't quite account for. But, on another, they were quite wise in the sense that they didn't want to trouble what was already working for them with any active reflection (or, perhaps more rightly, they didn't have the time or the inclination to do something different). Pat, a comparative novice who worked for a pharmaceutical firm, was reflective on this process. Particularly so because she had only been running for a couple of years now. She was keen on getting herself into that kind of state – getting into a positive habit such that things were no longer a matter of personal choice for her. In fact, she was 'afraid' that she'd get back into her old ways of doing things in which she'd 'find herself stopping' as she often did towards the start of her attempts at becoming 'a runner'. She wanted not to be thinking about her running. Indeed, she was actively aspiring to autopilot. And when you are on 'autopilot' speaking in terms of 'this is working for me at the moment' helps the system you'd worked hard to establish to stay untroubled by active deliberation about whether it could work better for you.

Others, however, were more reticent about talking about their running in such a register. This was, I think, partly an issue of self-presentation and partly a matter of how some felt that a research interview exchange should probably proceed. For example, one of our strategies (that we deployed as a means of pushing our respondents to reflect on what it was that they liked or what particularly 'worked for them' within their current running practice) was to pose a series of questions about how they might react if the context of their running was tinkered with in various ways. What if you could run outside without people being around? What if the gym had bigger windows overlooking a park? What if the weather was nice all the time? Admittedly, these were advanced towards the end of our time together and so, and as was the case in the following instance especially, many of the runners were getting a little tired by this point. Still, the degree to which both groups were reticent about these questions was, however, clear. They brushed aside the suggestion of amending their current run with quick answers that allowed them to rattle through these hypothetical situations (yes, I'd do this; no, I wouldn't do that). Again, what they were doing was pushing certain aspects of their current practice away from active reflection.

Tommy, who ran every day (more or less) in a downstairs gym at his workplace, had a particular way of dealing with being quizzed on these matters. He didn't like to position himself as being 'carried along' by his existing running practice. Rather he preferred to see this as an outcome of his intentions. He didn't 'need' to go running, but he 'liked' doing it – the difference was subtle, but it was important for him. This was about seeing himself as the master of his actions (if he 'needed to', he had no choice; if he merely 'liked it', then running was always available for him to pick from a suite of exercise options). When we asked about the various scenarios described above, he too was reticent about questioning whether things could work differently for him. His repeated answer when quizzed on his thoughts was that this 'was not my preference'. I think now that 'preference' was probably the wrong way of talking about it. Or, at least, if he was making an active choice, that had happened a while ago and he was no longer really comparing. This phrasing was about how he wanted to position his actions in talking together with us. Indeed, the fact that he decided to speak in terms of 'preferences' was almost as though he was mimicking a social survey, since they are much more likely to ask about what people would prefer (and then leave it at that). He was transforming our interaction into a kind of questionnaire in which he could mentally tick the preference box. In so doing, he could continue with a vision of him being fully in charge of his actions whilst, at the same time, sidestepping the idea of evaluation.

In another meeting in central London in a nondescript office at her work-place, Bronwyn, who regularly runs a couple of times around Regent's Park (stay-ing generally on the path because that felt easier, even though, like James and his incline, she had 'heard somewhere' that running on grass was better), had a similar take on the relationship between active choice and lived experience. In a reflective moment towards the end of our time together, she pondered: 'I feel that I should change my running more than I necessarily want to.' Why did she feel that she should? And why didn't she want to? This was partly about how she seemed to think that critical evaluation, and the perfecting of a practice, was val-ued both more generally and also in running circles. However, doing that would require effort. And she got different things 'out of it' for the moment. There was a background sense that active improvement was valued, but it was also a hassle to embark upon that process.

All the Chat

Both groups were asked whether they were 'runners' and both approached this mantle quite gingerly. This was partly explained by how they had little previous occasion to engage in any such definitional work since, as described earlier, many had established a way of running that worked for them and thereafter simply kept going. In that sense, it came as no surprise to see them sidestep evaluation by describing their experience as 'just going for a run', 'going to do a run on the treadmill' or, even more safely unspecific, 'just going to work out at the gym'. Yet this also belied a broader reluctance to consider the question of who was 'doing it right'. In this regard, their answers were much more than simple statements of fact. Rather conversations were being controlled in the heat of the moment so that context comparison was nudged off the agenda and running was reframed as a functional activity unworthy of significant reflection or discussion. So, they were all keen runners, but they weren't fully 'runners' (since to say that risked more evaluation than was wanted).

This reticence encouraged us to consider what (if such talk was sometimes approached with caution) kinds of exchange would be more welcome (and so more likely to happen) amongst runners outside our interviews. If spoken ex-changes on this matter were infrequent, and often carefully controlled, what kinds were allowed? This was a matter on which a few, tellingly male, outdoor runners spoke most eloquently. Tim, who was a keener runner than most (he had paid for some coaching in the past so that he could perform better in the

10K races that he sometimes entered), spoke of how some 'had all the chat'. All the chat, in this case, was taken to mean an ability and interest in reproducing a socially sanctioned way of speaking about running. And that meant talking of the times that had been personally clocked for specific distances, 'training' (as though everyone was perpetually preparing for events), and particular routes and performances. This was also seemingly about benchmarking your ability against the person you are speaking with and sharing tips ('someone told me once'?) about improving in the socially recognised way.

This kind of talk occasionally crept into some of our own discussions. Indeed, it could be tempting to initiate it at the start of an interview. If the tone of the encounter is still in the balance because two comparative strangers have yet to settle into the (albeit temporary) intimacy of an interview, it can be easy to grab at certain interactional modes that seem more familiar and accepted. And so, in more than one instance, conversations began with this kind of comparison. Indeed, sometimes conversations seemed naturally to start by comparing past experiences. Perhaps this was a failure on our part. Perhaps we should have worked harder to establish the right atmosphere for a different encounter (that was not about rehearsing the 'chat' that might otherwise 'naturally' happen). But another way of looking at it would be to see this as providing a sense of the talk that was most readily attached to running and, based on that, exploring the implications.

Keeping Comparison Away

On an indoor run with Pat, she is deciding which treadmills we should choose. Some faced a wall with television screens playing music videos. The others looked through a window onto the expanse of trees and lawn next to her building. She preferred the second set. Then she could 'pretend' she was running outside. Why did she want to do that? Her answer was that 'outside' was where 'the action' really was.

In saying this, Pat was unusual in connecting her existing running practice to beliefs about where running 'should' ideally happen. Both groups tended to recoil from too strong an evaluative stance on this topic. For the outdoor runner, the difficulty was partly about how they felt they should position themselves with regard to their indoor counterpart (were they doing something similar or different?). The attraction of the former framing was that it sat well with their desire to be seen as non-judgemental advocates of the exercise they personally enjoyed. In line with this, and as a means of avoiding any implied condemnation, some went as far as to paint the indoor runner as even more impressive in the sense that they kept going despite the imagined boredom of the treadmill. Others claimed that they had 'no idea' why they personally preferred running outdoors in a way that was partly about preventing any further exploration but also surprising when it was also clear elsewhere that they thought they were running in the superior environment – the place where, as Daisy told us, 'real running' happened. Maddy, for example, performed some deft conversational repair work by initially stating 'I just don't understand treadmill runners' before quickly adding 'but they are no different from me'. The outdoor runners were convinced that their environment was preferable, but they were also coy about saying so.

The indoor runner meanwhile was certainly amenable to a run outdoors in principle and, as already discussed, would likely position the outdoors as where recreational running ideally happened. Yet still they stayed inside. As a means of exploring this contradiction, we tabled a number of explanatory options in our interviews. These variously included concerns about safety, dirt, pollution and unwanted personal appraisal when running through public spaces. Yet these seldom struck a chord. Rather their concerns were more about personal control and predictable experience. For example, Simon's lunchtime run happened in the gym near his office even though he readily recognised how (to link back to what I learnt in the last set of interviews), 'in principle', passing through the parks and other areas of relative greenery not far from his office might be more pleasant. But, on reflection, and in actuality, his indoor run was somehow more 'relaxing'. After a short pause, this was identified as being because he could guarantee the indoor run would end just when he wanted. And that meant he could enjoy it untroubled by any worries about making it back to work on time. Were he to run outside, the weather might change or he might misjudge the time required by a particular route. These mental and physical distractions meant that he had 'no time for naturalness' as he joked whilst, in his conversation with us, he started to develop a narrative that prevented his own running environment from being seen as inferior.

In this respect, both groups tended to frame the competing environment (outdoor spaces for the indoor runners and vice versa) as likely to lead to unwanted levels of mental engagement. Though, of course, they were entirely happy for others to do their exercise there, and they wanted to support them in doing so, personally they just felt that the other environment might not be 'right'. So, the emphasis was very much on what 'personally' worked for individuals again. For the indoor runners, outdoor running was treated with some suspicion because it was taken to entail responding to various disruptions to do with weather and people and other physical features. For the outdoor runners, the experience of running on treadmills could be taken to be so monotonous that runners might soon start thinking about what they were doing in a way that might risk unsettling the habit.

In both cases, however, what these comparative assessments underscored was how active thought about certain features of their current running was unwelcome. These respondents had, in effect, set up what they felt was a positive practice and, because they did not want it destabilised, certain topics were pushed away from both contemplation and conversation. And neither did they want to question the environments in which others ran because they had set up a positive practice too (so it was mean to get close to condemning any part of how they did that). They were speaking in ways that were partly about admitting a lack of prior consideration as discussed above, but also about steering the discussion away from unappealing and uncomfortable topics that skirted the idea of judging those with whom they liked to see themselves as having a vague affinity. And so, to return to Simon, who elsewhere in our conversations struggled with the idea of being 'slave to a routine', ended up, as we considered comparison, saying his indoor treadmill was 'just where I am!'

Improving the Present Environment

Hans, a German lawyer, runs three times a week in the area around his East London home. During a rainy run along a usual route, he was happy to talk as he went. He was asked about the best features of that route. Though there were some scenic elements and a significant stretch along an attractive towpath, initially he didn't have a great deal to say. Some tentative opinions were offered about trees, traffic and views. Eventually, however, he admitted that this route does get boring. But, on reflection, he said he liked that it was 'reliable like that' – he knew where he was going and how long it would take – and that reliability helped to keep him running.

It may come as no surprise that both sets of respondents were relatively attached to their current running routes and routines – they were, after all, often running and so frequent conscious innovation would have been quite a challenge. They were also admittedly selected because they generally ran in one or the other environment. And they also had many other things going on in their lives – indeed, a good number were city professionals (so many of them embodied the 'office work' discussed in Chapter 3): as lawyers, as consultants, or in other professional roles as academics and office managers. Expecting them to devote a great deal of time to thinking through how change in their running could happen was, in some senses, quite an assumption. Indeed, as many of them were at pains to emphasise, they had many competing preoccupations (in line with the earlier market research finding that running appeals to the busy). And that made innovation less likely. Nonetheless we still sought to examine whether they might change how they ran (perhaps even where they ran). Yet, on tabling the topic of their personal running futures, responses were again lacklustre. Irrespective of the detail of how this was justified in the interview, we saw a distinct lack of conversational relish. Nonetheless, we did note how, in terms of how current environments could be 'improved', responses differed according to the group.

We might have imagined the outdoor runners would be more satisfied with their current running experience. Though (as described already) they had good reason to be reticent about saying so, they clearly thought they occupied the superior environment. Yet, in actuality, they had more to say about how that environment could be improved. Sometimes this was in terms of the volume of other people they encountered, with some clear opinions about the numbers that provide interest without becoming a navigational problem. There were also weather issues and quite a bit of lively talk about the difference made by certain surfaces. For the indoor runners, meanwhile, improvements were harder to identify. Again, this was partly because the point was not to reflect too much. But it was also because the gym provided a near ideal environment for unthinking exercise. Gyms were good, as Anya put it, for those not wanting to 'disturb things'. And so even though, in principle, the outdoor runners thought indoor running was boring and the indoor runners thought outdoor running was better, in practice, the treadmill runners were more content. When I attempted a joke about this being a 'nice day for a treadmill run' on a sunny day with Charlie at her workplace gym, her blank expression did not reassure me that I'd established rapport. Outdoor running was too removed from her current practice for it to register as a conversational candidate, let alone for it to occur to her to respond to my attempt at humour.

Oh, I Love the Outdoors!

Bronwyn is telling me about her favourite part of the run that she commonly does around Regent's Park (one of the biggest parks in central London with a variety of things going on within it – from a school to ponds, to a cricket pitch, along with plenty of lawns, paths and trees). This is when we are returning to the park's south east corner. Recently she was running across a certain expansive grassy area – there was nothing but her 'and the trees and the sky and the grass'. The sun was streaming down upon her and she loved it! She loved it partly because this was unusual (good weather, few people around) and partly because this was the point in her run when she was feeling 'energised' anyway (she was nearing the end but was still not too tired) and so felt she was susceptible to these kinds of feelings. Whilst she was on the topic, she also talked about how she was caught in the rain the other day. She liked that too. But in a different way. Her description was almost filmic – she saw the clouds gathering on arrival at Regent's Park and took a chance with the normal route. She got drenched, but kept going, and she 'actually loved that' in the end.

Hers was an almost moving account of being momentarily immersed in the elements. But the point to make here is that it stood out because it was also unusual. Most of the runners, both indoors and out, were sheepish about discussing such experiences. And it was easier to do when the discussion drifted (or was pushed) away from the experienced detail of their existing running practice. When we moved towards the benefits of outdoor environments towards the end of our time together, it was common to jump immediately to other activities (Oh, I always make time to be outside on the weekends! Yes, I love going to parks whenever I can!). They seemed more responsive to these urges in other domains of life.

Returning to Charlie, she, for example, responded to this topic with 'Of course, I love the outdoors!' – it was a self-evidently good thing is what she was effectively underlining. But she was also shutting down the conversation, discouraging further enquiry (why would I ask more questions about a self-evident thing?). Some of the outdoor runners were similarly uncommunicative on this topic. Kim said she 'liked trees'; Fiona thought that 'green parks were nice'. After moving away from 'the chat', Tim described Regent's Park in less poetic terms than Bronwyn. For him, it was a 'nice place' to run because there was 'plenty of green' (though it was still urban). He had much more to say, however, about pedestrians as obstacles. Geoff said the 'indoors was deeply uninspiring', though there was much less to say about how exactly the outdoors inspired. Bob said that green spaces allowed him to 'zone out' and left it at that in the discussion. Outdoor benefits were recognised. But they were not eagerly talked about by either group: being there was nice, but it had also faded into the background of their consciousness (and again that was irrespective of the group).

Far Away from the Outdoor Ideal

We asked all of the runners involved in this second study about their 'ideal London run'. In response, they painted various pictures of running in parks with the right number of people around or along the River Thames as shafts of sunlight repeatedly pierced the tree foliage above. Even though that was exactly where half of our respondents were found, the ideal run was never indoors. And this could partly be explained with reference to the above discussion of how, once established, respondents were reluctant to tinker with their existing running practice (despite a background sense this might take them closer to their ideal).

The point was that where they would like to run and where they ran were very different things. Or at least they were the starting point for very different mental trains of thought. There was a discernible shift in the pace and the style of their talk when we moved from the one to the other. One was about a nervous interrogation of something that many had good reason not to question. The other led to a much happier exchange as runners started to relax and sit back in their chairs and let their minds wander across the city, searching for the most attractive running environments, safe in the assumption that this hypothetical exercise was disconnected from a current practice that generally went unspoken and unexamined.

Conclusion

Good Reasons for Avoiding the Outdoors

Whilst recreational running may be one of the easiest exercise forms for the members of increasingly rushed societies to slot into their lives, and though running might be thought of as straightforward because we are partly designed to do it, the processes by which individuals come to end up running in particular places are another matter. It was with that in mind that this second case study set out to explore how speaking with recreational runners who generally did their running either indoors or out could provide an appreciation of that process and draw out some implications for the encouragement of positive greenspace experience. This was sometimes challenging since, compared to the office workers, certain aspects of their running were seemingly best undiscussed. And a key point in that respect was that this was something that only became clear to them during the interview in a way that could surprise them. Though these were often quite thoughtful people in other domains of life, in the interview they found themselves unwilling or unable to discuss something they were quite willing to consider by virtue of having signed up to the project – in some senses, a strange situation, but as we have seen, in others, quite understandable.

What are the implications? Returning briefly to the evolutionary perspective, what we found was rather perverse. For our indoor runners, it was perhaps exactly because running may be experienced as something people naturally do, that they find it so easy to stay on the treadmill. In other words, it is precisely because the action came so easily that any ideas about locational amendment were soon put aside in a further ironic twist in the tale of how modern humans are living with their evolutionary legacy. Returning briefly to the work on the benefits of green exercise, we saw how, in what might be understood as another perversity, though runners may be intuitively aware of the benefits associated with outdoor running, comparison was often off the cards because the point was partly to cede control to the practice. In other words, it was exactly because they wanted to keep going with their running that they didn't want to reflect on what environment was better and whether they were in the right one. Over the course of doing this study and afterwards, various interested parties have asked why these runners had 'chosen' their current environments. Our response was that this was the wrong way of characterising the process. Though the idea that action follows intention is enshrined in many accounts of social life (and challenging how often it applies has been one of the motivations for this book), these runners were rarely making conscious choices about how and where they wanted to run. Indeed, they had good reason not to give some aspects of their running much thought.

In this respect, I'd characterise this second relationship with the everyday outdoors as one of avoidance for the indoor runners. It was less a case of the idea of potentially going there gradually drifting away from any active consideration, as it commonly was with the office workers, and more a matter of how it contextually made sense to avoid this line of thinking. But in saying that, this was a subtle kind of avoidance. There was little sense here of runners deliberately dodging parks (determinedly turning away from the dangerous distractions and temptations of trees and plants!) as they rushed past on their way to the reassuring familiarity of the indoor treadmill. This was more about a kind of mental avoidance since engaging more fully with the idea of running outside would unsettle an otherwise valued practice that they wanted to persist in their lives. In terms of possible interventions, we cannot reasonably expect ideas about better environments to spread through the existing social networks of recreational runners (as though such networks always existed in the first place). Running was personal and there was good reason not to discuss it in this way. Nonetheless, when running reflection happens this can entrain a process of practice refinement and potential relocation.

Perhaps it may be better to catch people at the start of their running careers since otherwise they will settle into whatever environment they have found themselves. And that may be especially easy inside the gym because of the absence of environmental variations that could trigger thoughts about how and where the runner would ideally run. In other words, such variations could serve to unsettle a practice that was, even if it happened in a less than perfect environment, understandably valued. And so we can easily imagine a situation coming to pass similar to that depicted on the front of this book. Whilst, on one level, it does seem odd to see someone running on a treadmill that faces out onto vegetation through a window (if you think the outdoors is the better environment, why don't you just go there?), now we can see how doing so also makes a lot of sense. Greenspace experience is good. But it is also good not to jeopardise the benefits of exercise by threatening the existing practice with too much reflection. In other words, by taking the approach that I did (looking for the subtleties of how and when practices repelled thoughts of doing differently), I could see the scale of the challenge for those who might want to encourage runners into alternative environments.

Stopping to Talk When You Are Up and Running

Returning to the second wager, how did my experiences in this second project relate to my three suggestions? What did the runners tell me about the responses and reactions that may be worth looking for when hoping to learn from how people speak about their practices?

- The first suggestion was that, in order for practices to gain strength as activities, they often call forth forms of talk that help them to solidify and become more clearly recognisable activities out of a world full of potentially chaotic, and hard to fathom, human actions. In this second case, what was most interesting was how people were aware of, and sometimes actively negotiating, this process themselves. This was not really a case of this second set of respondents being unthinkingly drawn into particular ways of speaking about their running. The only obvious example of this in the runner study related to how people handled the idea of 'having all the chat'. Some took pride in having the chat. And certainly 'the chat' could make our conversations easier, allowing them pick up speed as those involved slotted into more established and familiar forms of spoken exchange. Indeed, during these conversations, on occasion, they jockeyed for position in terms of showing themselves to be more fully recruited into a practice they had, after all, often worked hard to embrace. Exchanging tips and ideas and anecdotes about performance according to the 'received wisdom' of running could also draw them more fully into it and might, as a consequence of that, eventually make for better running (as conventionally understood). Yet, when others alluded to 'the chat' in saying things like 'someone told me once', there was also significant ambivalence about getting involved in this kind of talk. Perhaps especially for such geographically dispersed practices as recreational running (in which runners often have little occasion to talk to one another about what they do), people may recognise the value of engaging with this kind of talk, but not be especially determined to seek it out. They were 'up and running' already and so had little need to stop and speak about it. So, in some cases, those involved have good reason not to (and, because of how it is practically done, also have little opportunity to) help the practice solidify in this way.

- The second was about attending to how people handled the experience. There were definitely some issues to be faced here. But those should be thought of as insights as much as obstacles. This was at times in a similar way to how some office workers occasionally balked at the idea of being 'carried along' by the practice (when they preferred to think of themselves in different terms). But equally many of the runners were quite ready to admit that they were not especially mindful of how their running went on. The best example was the breezy admission that treadmills 'must be bad' because they were not being used. This, in fact, provided me with a sense of the extent to which they wanted to embark upon this kind of evaluative exercise. And that, in fact, led to one of the crucial findings of the study, namely that, when you are doing something that you think is good for you, becoming critically self-aware of certain elements might not be all that attractive as a proposition. Of course, another, almost comedic, picture of what was happening in this second project was that we had pestered people into taking part in our study and, as a direct response to this, they revealed their unhappiness about this coercion by being deliberately unwilling to think too hard about their running lives with us. But, in actuality, once these runners had committed to taking part, and once they had allocated the time to do so, they tried their best. And so these aspects could surprise them as much as us.

- Third, what was instructive in terms of how these runners connected to the idea of the 'done thing' was how often they positioned themselves as personally apart from this idea. Indeed, the conversational manoeuvres involved in connecting and disconnecting from the idea of running as a collectively shared practice were especially revealing. There was often a clear separation between 'what should be done' and 'what I do myself'. And, unlike the office workers who were often doing much the same as one another (yet could nonetheless be quite keen on standing outside of the practice to diagnose the 'herd mentality' of their professional peers), in this case this had implications in terms of how a diversity of current running forms was able to persist. In other words, 'the ideal' was exactly that – a situation apart from what currently happened for them. And, because of this distancing work, a constellation of current practices was able to persist (if they can really be called practices when they are so personally idiosyncratic as this). Ideas about the socially sanctioned means of undertaking an activity do not necessarily draw people towards comparable actions. Sometimes these ideas can be pushed away as an unobtainable prospect for those who would rather do what suits them at the time. So we might, at times, question the assumed 'social' nature of practices (or rather treat their apparently 'shared' character as a topic for investigations that could involve talk).

I have now told a second story of unsettling outdoor environments. And we have started to understand why some are stuck on treadmills regardless of where they'd ideally run. We might have imagined that, during their leisure time, people might feel themselves to have more autonomy over their actions and therefore be more responsive to the benefits of outdoor experience. But, if anything, indoor attachment seemed to be stronger amongst those on the treadmill than those in the office. So, we cannot really presume. It is with that point in mind that attention now turns to the domestic garden. The British are often represented as devoted to their gardens. My next chapter starts with whether this remains the case.

References

Akers, A., Barton, J., Cossey, R., et al. (2012). Visual color perception in green exrcise: positive effects on mood and perceived exertion. *Environmental Science and Technology* 46: 8661–8666.

Bamberg, J., Hitchings, R., and Latham, A. (2018). Enriching green exercise research. *Landscape and Urban Planning* 178: 270–275.

Boullosa, D., Abreu, L., Varela-Sanz, A. et al. (2013). Do Olympic athletes train as it in the Paleolithic era? *Sports Medicine* 43: 909–917.

Burton, N., Khan, A., and Brown, W. (2012). How, where and with whom? Physical activity context preferences of three adult groups at risk of inactivity. *British Journal of Sports Medicine* 46: 11–25.

Butryn, T. and Masucci, M. (2009). Traversing the matrix: cyborg athletes, technology, and the environment. *Journal of Sport and Social Issues* 19: 423–450.

Carrier, D. (1984). The energetic paradox of human running and hominid evolution. *Current Anthropology* 25: 483–495.

Depledge, M., Stone, R., and Bird, W. (2011). Can natural and virtual environments be used to promote improved human health and wellbeing? *Environmental Science and Technology* 45: 4660–4665.

Eichberg, H. (1986). The enclosure of the body – on the historical relativity of 'health', 'nature' and the environment of sport. *Journal of Contemporary History* 21: 99–121.

England Athletics. (2013). *A Nation That Runs: A Recreational Running and Athletics Plan for England 2013–2017*. Birmingham: England Athletics.

Gladwell, V., Brown, D., Wood, C. et al. (2013). The great outdoors: how a green exercise environment can benefit all. *Physiology and Medicine* 2: 1–7.

Henley Centre. (2005). *Demand for Outdoor Recreation: A Report for Natural England's Outdoor Recreation Strategy*. Reading: Henley Centre.

Hitchings, R. and Latham, A. (2017). How 'social' is recreational running? Findings from a qualitative study in London and implications for public health promotion. *Health and Place* 46: 337–343.

Howe, P. and Morris, C. (2008). An exploration of the co-production of performance running bodies and natures within 'running taskscapes'. *Sport and Social Issues* 33: 308–330.

IBIS. (2018). *Gyms and Fitness Centres in the UK*. London: IBIS.

IBIS. (2019). *Sports Facilities*. London: IBIS.

Jungers, W. (2010). Barefoot running strikes back. *Nature* 463: 433–434.

Keynote. (2010). *Sports Market*. London: Keynote.

Keynote. (2013). *Health Clubs and Leisure Centres Market Report*. London: Keynote.

Liebenberg, L. (2006). Persistence hunting by modern hunter-gatherers. *Current Anthropology* 47: 1017–1025.

Lieberman, D. (2013). *The Story of the Human Body: Evolution, Health and Disease*. London: Allen.

Mackay, G. and Neill, J. (2010). The effect of 'green exercise' on state anxiety and the role of exercise duration, and greenness: a quasi-experimental study. *Psychology of Sport and Exercise* 11: 238–245.

Malina, R. and Little, R. (2008). Physical activity: the present in the context of the past. *American Journal of Human Biology* 20: 373–391.

Mattson, M. (2012). Evolutionary aspects of human exercise – born to run purposefully. *Ageing Research Reviews* 11: 347–352.

McDougall, C. (2010). *Born to Run: The Hidden Tribe, the Ultra-Runners, and the Greatest Race the World Has Never Seen*. London: Profile.

MINTEL. (2010). *Sport and Exercise: Ten-year Trends*. London: MINTEL.

MINTEL. (2013). *Sports Good Retailing*. London: MINTEL.

MINTEL. (2018). *Sports Participation*. London: MINTEL.

O'Keefe, J., Vogel, R., Lavie, C. et al. (2010). Organic fitness: physical activity consistent with our hunter-gatherer heritage. *Physician and Sports Medicine* 4: 1–8.

O'Keefe, J., Vogel, R., Lavie, C. et al. (2011). Exercise like a hunter gatherer: a prescription for organic physical fitness. *Progress in Cardiovascular Diseases* 53: 471–479.

Office for National Statistics. (2011). *Lifestyle and Social Trends*. London: HMSO.

Park, B., Tsunetsugu, Y., Kasetani, T. et al. (2010). The physiological effects of Shinrin-yoku (taking in the forest atmosphere or forest bathing): evidence from field experiments in 24 forests across Japan. *Environmental Health and Preventative Medicine* 15: 18–26.

Plante, T., Cage, C., Clements, S. et al. (2006). Psychological benefits of exercise paired with virtual reality. *International Journal of Stress Management* 13: 108–117.

Pretty, J., Peacock, J., Hine, R. et al. (2007). Green exercise in the UK countryside: effects on health and psychological well-being, and implications for policy and planning. *Journal of Environmental Planning and Management* 50: 211–231.

Pretty, J., Peacock, J., Sellens, M. et al. (2005). The mental and physical health outcomes of green exercise. *International Journal of Environmental Health Research* 15: 319–337.

Raichlen, D., Armstrong, H., and Lieberman, D. (2011). Calcaneus length determines running economy: implications for endurance running performance in modern humans and Neanderthals. *Journal of Human Evolution* 60: 299–308.

Tam, N., Astephen Wilson, J., Noakes, T. et al. (2013). Barefoot running: an evaluation of current hypothesis, future research and clinical applications. *British Journal of Sports Medicine* 48: 349–355.

Thompson Coon, J., Boddy, K., Stein, K. et al. (2011). Does participating in physical activity in outdoor natural environments have a greater effect on physical and mental wellbeing than physical activity indoors: a systematic review. *Environmental Science and Technology* 45: 1761–1772.

Ward, M. (2017). Swimming in a contained space: understanding the experience of indoor lap swimmers. *Health and Place* 46: 315–321.

Chapter Five
Succumbing to the Outdoors: In the Garden

A Nation of Gardeners?

In 2018, 58% of Britons said that 'gardening is a real hobby for me' (Global Data 2018). That may be just as well when 70% of homes in the United Kingdom have gardens, as a country with around 20 million domestic gardens (MINTEL 2018a). According to some estimates, this is around 3–4% of the total landcover, representing between 35 and 47% of urban greenspace in the United Kingdom (see Cameron et al. 2012). We also know that, according to most reports, the market for gardening products, which was already large, continues to grow (MINTEL 2018a; IBIS 2019a). All this would initially suggest that Britons continue to enjoy a close relationship with their gardens – that their future is, so to speak, rosy. We might have expected this when the cultural idea that Britons love growing plants at home was already well established (Uglow 2017).

However, were we to look a little more closely at some of these reports, we might start to tell a more sobering story of disconnection. First, we know that demand for space in an already crowded country means that gardens are shrinking (Global Data 2018). It also emerges that the garden market is partly propped up by an ageing population as people 'get into gardening' in later life because they have the time then and more moderate levels of physical exertion have started to appeal (Global Data 2018; IBIS 2019b). Younger people, those in cities and those

The Unsettling Outdoors: Environmental Estrangement in Everyday Life, First Edition. Russell Hitchings.
© 2021 Royal Geographical Society (with the Institute of British Geographers). Published 2021 by John Wiley & Sons Ltd.

who are more affluent have a more uneasy garden relationship, it seems. According to MINTEL (2018a), these were most likely to say that garden centres were 'full of confident gardeners' as though they personally found them intimidating. A closer look at what is being bought also hints at a decoupling of people and plants. The biggest 'garden spend' is now garden furniture (MINTEL 2018a), which is nothing at all to do with growing plants. All this is apparently in line with a wider trend towards seeing the domestic garden as a place for passive contemplation more than active cultivation.

Britons are now, for example, more likely to say that 'nothing beats relaxing in the open air' when they are asked to define their relationship with their domestic gardens than they are straightforwardly to say that 'gardening is enjoyable' (MINTEL 2018a). Then, turning to the tools that they use there, power equipment increasingly seems to appeal since the sales of such items, compared to their non-powered (perhaps more unattractively labour intensive) equivalents, has grown at five times the rate. This could, of course, partly link to how older gardeners may have less energy. But the trend still remains. And it suggests a desire to 'get things done' more than a process of regular tending. Equally, in recent years, when it comes to the plants themselves, the lowest rise in sales is perhaps tellingly for 'seeds and bulbs'. This also suggests that many Britons no longer want to wait too long for things to grow. Indeed, some reports see the 'confident gardener' being replaced by a kind of 'convenience gardener' who prefers to get their 'outdoor room' efficiently organised. Trends such as these encourage Ginn (2017) to see a 'hollowing out of gardening culture' in the United Kingdom.

In line with my particular focus in this book, there are also some interesting ways in which people orientate themselves towards the garden under the circumstances of being asked about their relationship with it. If we go briefly back to the poll that started this chapter (in which 56% of Britons said that 'gardening is a real hobby for me'), the same poll also found exactly the same percentage agreeing that 'when I garden I want to see the final effects quickly rather than have to wait for it all to grow' (Global Data 2018). They also found a greater number of people (65%) saying 'I expect to spend more time in my garden in the next 12 months than the last 12.' So, many Britons like to, at least, think of themselves as those who enjoy gardening. And they like to think of themselves doing more in future than they presently do. But they also like to get things done quickly (which doesn't quite seem to fit with the suggestion of an ongoing and absorbing garden engagement). There are tensions here between getting things done quickly and growing a garden – tensions that might be seen in how we shop and whether we 'garden'. But, before we get to the detail of how these tensions are negotiated in practice, what of the greenspace benefits that gardens provide?

Gardens as Both Beneficial and Bewildering

In terms of the beneficial effects of gardening, the evidence is overwhelming for some. One recent review of existing studies on this topic, for example, concludes that gardening can evidently do all sorts of brilliant things to us (Soga, Gaston, and Yamaura 2017). And that is why this activity has attracted a growing amount of research attention. Gardens, it would seem, could help us to solve various societal problems that go beyond psychological restoration – though that is certainly something that being around garden plants can seemingly provide. These include the physical benefits that come from the exercise associated with tending to plants and keeping on top of what is happening in the growing garden, the social cohesion and support that can come from gardening alongside others who you might not otherwise know, and the sense of personal achievement that can come from helping the garden to take the form that you hoped it would assume. In summary, Soga, Gaston, and Yamaura (2017, p. 92) are happy to say that a 'regular dose of gardening is good for public health' (see also Thompson 2018).

But there are issues, as there always are, with thinking in particular ways. One of the concerns that some have about the above account is that it glosses over how things might work out differently for different social groups. Partly because of the hopes that have been invested in the potential power of gardening by those who have undertaken work on this topic, many studies have focused on those who might particularly concern us in public health terms. In line with this ambition, there have been studies focused on the benefits of gardens for those in later life (when domestic gardens are often comparatively easy to access and might provide exactly the regular kind of low-level exercise that ageing societies need) (Ottosson and Grahn 2005; Bhatti 2006; Scott, Masser, and Pachana 2015). We have also seen a good amount of understandable interest in those who are currently facing various mental and physical health challenges since the garden seems to provide a path to restoration that might be particularly good for them to take (Pretty 2004; Söderback, Söderström, and Schälander 2004; Parr 2007). But, if we focus only on certain social groups, we learn comparatively little about the 'everyman and everywoman' of current garden ownership in terms of the likelihood of these benefits flowing out of their gardens to them too (along with the challenges that they face in getting to a point when they are capable of deriving them).

This takes us to another issue that is understandably often downplayed in studies of 'therapeutic horticulture' (that encourage gardening because of the benefits), namely the question that the 'everyman and everywoman' of garden ownership immediately face after moving home: what am I supposed to do with it (see Taylor 2008, on this)? People don't just magically start gardening. Indeed, 'having a garden' and 'doing gardening' are two potentially very different things. And assuming that one leads to the other is quite a leap. What is contained with this broad-brush idea of 'gardening' in any case (Does sweeping the decking count? What about garden furniture repair?)? Gross and Lane (2007) have considered how gardeners somehow get 'hooked' on an activity that involves tending to plants. But doing so is far from a done deal. Pitt (2014) wisely points to how the pleasure of gardening is about learning to go with the 'flow' of how life goes on within these spaces. But getting ourselves into this kind of trance-like state might not always be so easy. And, following some of the arguments that I developed earlier, whether the 'flow' gets going (and whether people get 'hooked') is partly down to whether certain ways of thinking and talking about the living creatures that might feasibly find a home in our gardens take hold.

In this respect, like the Danish office workers at the start of Chapter 3 (who blithely said there was 'no reason' why they didn't go into the greenspaces near their work) and in a similar way to how some of the treadmill runners from Chapter 4 also proved quite determined to disengage from certain questions, we could think more about what particular ways of speaking around domestic gardens reveal. For example, Gross and Lane (2007) emphasise how eager potential research subjects can be to talk about these topics. But it has also been noted how getting people to talk about gardens can be enjoyable for them because they hadn't thought a great deal about the matter before (or, perhaps more accurately, their ways of thinking about gardens didn't match up to those determined by the researcher's questions) (Freeman et al. 2012). Then there are also hints of how respondents might find it difficult to talk in certain ways about their gardens – for example, in terms of how some of them might brush certain forms of garden benefit analysis aside with 'it's a psychological thing' (Freeman et al. 2012, p. 139). Building on these suggestions, this chapter will consider some of the forms of talk that take place around the 'domestic gardens' of London (rather than seeing what people are inclined to say about what 'gardening' gives them and taking that as a straightforward expression of their feelings).

Taylor suggested in 2002 that TV gardening experts had become quite reticent about positioning themselves as such. Perhaps being reminded that effective gardening requires knowledge that takes time to acquire was not what the viewers wanted to hear when watching shows focused on how great gardens could be quickly created. After all, the lived reality of being presented with a collection of strangely unfamiliar living creatures outside your home, and which you are often expected to have 'under control' (Head and Muir 2007) can be a daunting prospect. Certainly, it won't straightforwardly lead to a suite of wellbeing benefits. As Tims (cited in Freeman et al. 2012) sagely reminds us, 'gardens are frail and ephemeral things, utterly dependent on the dedication of their carers and always teetering on the edge of ruin' (2006, p. 2). That sounds stressful. And perhaps it is especially so for those more familiar with living in line with some very different logics.

Shopping versus Gardening

This way of thinking encourages us to see living plants as a problem for those more used to dealing with material goods without thinking so much about how their physical properties might complicate the experience. It also takes me to the practice to which I want to attach this chapter – shopping. You might have expected a focus on gardening. And there will be some discussion of how the idea of 'gardening' is handled by those who concern themselves with London domestic gardens. But my interest is also in how certain common social practices could be serving to shape ongoing relationships with outdoor spaces in everyday life. And shopping, for authors like Zukin (2005), is definitely common and could even be said to define many of us now. Certainly, it takes up a lot of our time in the United Kingdom (Gatersleben et al. 2018). And so, whilst many may recoil from the idea of thinking of themselves as 'shoppers' (in view of the image that is often conjured up of a compulsive consumer who delights in the transient pleasures of taking ownership of possessions), the practice of shopping is still woven into the fabric of many societies today. It is, after all, fundamental to how we commonly define our collective health – when people are buying more goods (when 'consumer confidence' is up) the idea is that things are going well.

How people go about selecting, acquiring and then living with consumer goods has also been changing in ways that are relevant here. The tools needed to shop effectively, for example, have mutated from cars and bags to laptops, delivery addresses and smartphones. As Cass and Shove (2017) point out, whilst there are many ways of shopping, the internet is playing a growing part in this practice. This begs the question of how this shift is shaping how shoppers relate to the physicality of the items that they buy. For example, younger people in the United Kingdom are more willing to buy clothes online (MINTEL 2018b). Older generations, by contrast, more often like to try items on first. They are also almost twice as likely to say that they wouldn't buy from websites because, without this kind of direct experience, they are unsure of the 'quality' of what they are buying. This perhaps partly explains the rise of 'fast fashion' – the idea that young people especially are buying more clothes of cheaper quality partly because they know that they will soon throw them away. In this kind of shopping system, 'quality' (in terms of durability) matters little because the purchased items will likely be discarded long before their longevity is put to the test (Bhardwaj and Fairhurst 2010). Some new ways of relating to the materiality of goods can be discerned here.

If this kind of physicality is often backgrounded by increasingly widespread shopping practices that are about immediate visual impact (and downplaying the demands that our purchases might make of us in terms of care and maintenance after acquisition), what are the implications for the domestic garden? In some ways, this returns us to the 'garden on the edge' idea (Freeman et al. 2012) that was introduced above – the idea that the domestic garden is, far from the sanctu-ary from the modern world that we may romantically like to imagine it as, full of potential faults and challenging, upsetting and unsettling activity. 'Gardeners' may love plant liveliness. But whether contemporary shoppers do is another mat-ter. In this respect, online shopping for plants might, for some, still feel rather odd (like getting pets sent through the post?). But, for others who are alive to new business opportunities, it is an expanding market that just makes things easier. If efficient delivery gets 'the product' to the 'customers' before it starts to suffer, what's the problem (Baker, Boyer, and Hikaru 2018)? It would also seem that this process might be especially appealing for urban Britons who prioritise the efficient organisation of an attractive landscape. For example, it has recently been observed that those who live outside British cities have a different seasonal rhythm when 'getting their gardens ready' for summer (in itself an interesting idea in the suggestion that owners should work towards it, rather than wait and see what summer brings). Outside our cities, people do this earlier (Masnaghetti 2016) and are more likely to prepare in advance. This is unlike their, appar-ently busier and more impatient, urban cousins, who are more inclined to get the garden 'done' as summer approaches and less likely to possess the composure required to wait for seeds to work.

Getting into the London Garden

The above discussion sets the scene for my study of domestic gardens in London. In London there are still a lot of gardens, just as there are in many other cities around the United Kingdom (Cameron et al. 2012). But, also in line with the wider national profile, the amount of garden greenspace has been declining there (Smith 2010). That will be due to many things that include the desire to pave over these spaces so that cars can be parked. But it is also partly about the processes by which Londoners end up relating to these spaces. As I have now established, shopping is a practice that may be coming to influence UK domestic gardens. I now want to consider how this suggestion can be explored by paying attention to how some of the people involved think and talk about how best to handle these endangered greenspaces. They may like the idea of growing in principle, but, in practice, perhaps they now find it unsettling?

This chapter draws on a range of activities to explore how this suggestion was kicking into some London gardens between 2001 and 2003. People were inter-ested in gardens, I figured. So, I was hopeful that the task of researching them might be easier than with other social groups. My thinking was that they might be excited to talk to me about an exciting activity (more perhaps than the office worker would be about embarking on a dispiriting discussion about how seldom they ventured outside or the treadmill runner who had good reason not to think too much about where their running took place). However, I was also interested in those who owned gardens but didn't have too much time to learn about how to handle them. So, it was important not to assume that everyone would want to talk freely about what happens within these spaces (and also to look out for the times when they were disinclined, for various reasons, to say too much about certain aspects). The place in which an enjoyable hobby happens for one person might be a burden to another, after all, and that would likely be reflected in how they handled their spoken interactions with me.

One group that I wanted to spend time with were garden designers. A garden designer is someone a person would approach if they had an outdoor area at home that they wanted some guidance about developing towards what might be called a 'garden' (though exactly what a 'garden' is can be unclear in advance and the amount of living vegetation required is often, as we will see, something that emerges from the process). They could turn to friends, to neighbours or to other known gardeners for advice and ideas. But it is this group that makes it their business to step in at this point. Typically, a prospective client is visited on one or two occasions. The aim at this time is to establish what the client wants and to build rapport with someone who moments earlier was a stranger. Then, if the designer succeeds in winning the client's trust and business, they gradually come to a point when they supply some sort of 'design'. This is either in the form of a drawing or a more formal plan. Then, if the client likes it, it gets 'built', as they attempt to translate fantasies of future living and anxieties about practical management into a suitable physical form. And then they leave, hopeful that the client will take some pleasure from what they have provided (though some keep in touch afterwards). What, however, is especially interesting to me now in this chapter is how designers are forced to navigate two of the garden framings presented above. The first is the idea of a harmonious gardening co-existence with expanses of flourishing greenspace. The second is about shopping for the 'desired product' as efficiently as possible. How exactly do they tread a line between them?

I got to know some of these designers through a number of routes. I met my first designer, Kate, at a 'garden show' in north London. A second group came from attending some of the meetings in which designers came together to discuss their work. As I got to know some designers better, they suggested further designers as potential respondents. Two came through a third route, namely emailing directly through their websites. Each designer was interviewed twice, generally in their studios where they did their design work. Then I took a closer look at how they spoke with their clients by shadowing some of them on their consultations. After briefly introducing myself, I was generally silent through this process as I watched and listened to their discussions. The role I was often assumed to occupy was of an apprentice designer as I faded into the background. I also wanted to understand the client experience in a bit more depth. So, I asked some of the designers to put me in touch with some of their past clients. These people were, perhaps unsurprisingly, uniformly what would be called 'professionals'. They worked in a variety of white-collar fields including consultancy, marketing, estate agency, journalism, finance and art marketing. I went to their homes twice. We talked about how they lived with their gardens. Sometimes they'd offer a glass of wine. Usually we'd also walk around the garden and I'd take some photos. In our interviews, I particularly focused on how they experienced this process (and with what implications for the forms and types of actively living plants that ended up outside their homes). In total, I undertook 14 interviews with garden designers and 14 interviews with those who had hired one. Finally, in this chapter, I will also draw on some of the time I spent at some London garden centres. These are places where people go to buy plants and materials as perhaps a more obvious form of garden 'shopping'. In these centres, my initial approach was straightforward: I contacted all of those that I could find on internet search engines. Eventually, after some persuading, seven managers agreed to talk with me and I spoke with each of them twice. I also spent time within two garden centres, walking around and observing what customers did and said during their time there.

What Silence Suggests

I am strolling around a North London garden centre. It is a mild and sunny spring day. It is also Saturday, so there is plenty going on. Much more than during my mid-week visits when I could feel rather silly as I waited for 'something to happen' and chatted 'casually' (though appearing to do that was actually quite stressful) with the staff there. This busyness meant that I had to make some decisions about what to observe. I turned to a young couple in their mid-twenties. They had embarked on 'the usual lap' that I often saw – passing plants grouped according to type at the front, moving past the pots, and ending up alongside the landscaping materials at the back. I looked like a keen gardener with my field notebook. But there was actually a good number of gardening note-takers that came (writing down what was on sale before going back to see what they could accommodate at home). So in this instance I'm less worried about my jottings influencing people's actions. Their conversation flows as follows:

> 'It's ok, mmm, well maybe I'd quite like a bigger one, but I don't know,' she says. Then they go back into 'ums' and 'ahs' that seem to be communicating little. In the end, they put the plant they had previously picked back down … There are building works behind the centre and they talk animatedly about that for a while. It seems like something for them to latch onto – 'wonder what it's going to be' etc. Then it's back to comparative silence … Finally, the pace picks up: 'ah look at the cactus – they are great aren't they?' 'I love all the different shapes!' 'Shall we get some of those? I like that one, the big upright one.'

So far, so unremarkable. But there was a difference between the talk patterns prompted by plants and those that happened around things that were less alive. And I particularly saw this change of tone and pace from the younger visitors. Lyn, one of the managers with whom I spoke, had noticed this too. Indeed, she was increasingly alive to the business implications of these different responses in her centre. They meant that she didn't worry so much now about 'getting things right with the plants' because customers were easy to 'persuade' about those. She could give them whatever was available. They just wanted 'some plants', after all. However, she was quite sure they'd be much more resistant if she attempted to persuade them about ornamental features. As she underlined, you'd be able to 'palm them off' with a 'decent looking plant'. But you absolutely couldn't say they'd do equally well with a decorative metal chicken as one of 'those little stone Buddhas' (two of the options from her expanding range of ornaments). In particular, she, and others, felt that most 'central London' customers came with a strong sense of personal taste about these features. With the plants, however, they were more unsure. And that meant that she could give them some 'green things' that would 'just grow and they'd be happy'.

Elsewhere in a smallish garden in South East London, I'm at my second interview with Liz, an advertising manager whose garden was 'designed' and 'built' three years earlier. As I did with most of the 'designed garden owners', I encouraged her to walk me around the garden and, as we went, I asked what I should photograph for my study. The idea was that what they chose, and how they reacted, might say something about how they personally engaged with their gardens. We descended from the raised decked area and ventured onto the lawn. And there she waits. I was hoping she'd have some immediate suggestions. But this time I was faced with an awkward silence. She had paid a lot of money for someone to help her 'create the vision' of the garden that she wanted (and I had expected her to have a lot to say about the result). But no. I was left feeling silly again about a situation that was unexpected (it had worked well in other gardens). Later, she admitted that she also found it hard to know what to say. She definitely had more to say about how 'others' wanted beautiful gardens as a means of 'escaping the pressures of modern life' by developing a slower, more engaged relationship with living plants. In terms of wider social commentary (what garden design says about society) she was adept. Yet she had little to say inside her own.

Both of these examples speak of what the absence of speech around plants can (sometimes but not always, as we will see) suggest. Though the garden owners in this study definitely liked the idea of being around plants, though they were potentially drawn to them in line with the idea of a persistent innate attraction, still they didn't quite know what to say when they got there. And this had implications in terms of the gardens that some eventually ended up with. So, silence, in this first set of cases at least, suggested a kind of bemused inability. Plants did not slot into more familiar modes of spoken evaluation (with me or with others). Plants were uninterestingly 'green things' from which it could be easy to disengage.

Discussing Design with 'Central London' People

There was also (as seen in the above discussions with Lyn and in how some of the designers spoke) an apparent variation to this reticence – a variegated urban geography to the inclination of Londoners to speak in certain ways about plants. And, irrespective of the truth of the matter, the fact that they had such perceptions could mean that meetings were managed differently according to where people were. Beliefs about the 'central London' person could serve as something of a self-fulfilling prophesy (since they meant that certain groups were provided with certain sorts of gardens as a consequence). The idea was that 'central London people' had less free time, were less likely to tend to plants and wanted something stylish. We might have imagined them to be particularly eager, in principle, to take some time to relax and take care of their gardens in response to the apparent rush of inner-city life. However, the received wisdom was that the opposite was true.

But perhaps it would be more accurate to say that they were taken to be less expressive about plants in ways that made other conversational topics comparatively attractive. As I mentioned, when a designer meets a client to talk about how they might respond to the possibilities presented by the outside 'space' they had either acquired after moving or wanted to improve, a lot is up for grabs. That was especially so when clients were often taken to find it hard to discuss their apparent 'needs'. Certainly, this was taken to be the case by Margaret, a confident garden designer of long standing with whom I spoke in her basement studio in a leafy North London suburb. I imagined she'd be quite adept at drawing out what her clients wanted. But, by the end of the process, she (almost proudly) revealed that 'they end up talking and thinking like me' because of the structure to the encounter that she had imposed. She did so because 'they are generally not very good at articulating what they want' and so she had a 'responsibility to take them in the right direction'.

This was why Margaret coaxed her clients into thinking (and talking) about gardens in 'her way'. She did this partly so that they knew not to be scared of the garden (as she definitely wasn't). But there were also risks to embarking on this kind of imposition. After all, the designer also has to establish rapport and, if they don't immediately entice and delight with creative ideas and 'solutions', then the client might easily decide against hiring them. One strategy was to indicate how you'd already a good sense of their 'style' by looking at how they had decorated their homes indoors. Taking a cue from clients' interior design choices and showing how they could be mirrored outside was a common strategy, irrespective of whether clients might want something very different from the garden. This process helped me to understand the popularity of the 'outdoor room' idea about which many designers talked (often with ambivalence). Thinking, and talking, in this way was partly an indirect consequence of how empathy needed to be efficiently established in the context of a nervousness about whether clients could speak about plants as actively living things.

Another strategy was to find ways of helping clients to take existing activities (that they already knew how to do, and that they enjoyed doing, indoors) out into their gardens (for example, dining or having coffee with friends, playing with the children). Geoff, who I met with his design partner in a temporary office that provided them with a base in East London, considered this strategy over a cup of tea. He pulled out the 'tick list' that he always used 'to get them going'. On it were listed a series of hard landscaping features and activities that potential clients might want to think about relocating outside. One of the reasons why, in his opinion, the tick list was a good idea was because it ensured there was a structure to their interaction that made him seem like an efficient professional. But it also meant clients were less likely to think of further ideas about what they wanted to do in their gardens afterwards. And particular plants (and indeed 'plants' more generally) were notably absent from the list (he would select those afterwards, according to his professional judgement rather than their personal preference). These were gardens in which, for understandable interactional reasons, plants went undiscussed at the start because they were either uninteresting or unsettling to clients who should be kept comfortable. Both strategies, of course, pushed the idea of gardens potentially being unlike 'outdoor rooms' off the table. And so, in different ways, the templates of interaction used by these designers could serve to make the garden either a comparatively inert thing or a mere backdrop to activities.

Plants Guaranteed

Talking of plant care and maintenance with customers who were perceived to be nervous about these activities was an issue in the garden centre too. According to Dave, the manager of a West London centre, people (especially the young, wealthier people that he thought he should focus on) often recoiled from the perceived 'mystique' of gardening:

> 'They think, you know, it's just this sort of mystical thing ... there are all these sorts of old wives remedies and you know you have got to do this now, you have to do that at this time, and don't plant this until there is a full moon and all this sort of malarkey, and, you know, they do get a little bit worried about it.'

His view was that his customers were used to potential purchases being more docile than this. And that was both when his staff spoke with them and in practice afterwards. Indeed, the managers had various anecdotes about how these issues kicked in at their centres. Dave had received complaints when some plants were wet (such that one customer got water on her bags) – they expected a particular kind of 'retail experience' and their usual purchases were anything but wet. Lyn had received complaints about plants with damaged leaves. Viewed in one way, that would, of course, make them a 'faulty product'.

Faced with a sense of impatience and an apparent lack of knowledge, one understandable response was to introduce people to plants that were especially inert. Herbaceous plants, for instance, that die back to the ground in the winter were, for example, considered a challenge to this framework. Out of season, all the customer would see on encountering these plants at the centre was a strange 'bucket of earth' according to Thomas. And his thinking was that it was unlikely now, first, that they felt they had the skill to coax something out of the bucket and, second, that they'd want to make the effort. Enter the 'plant guarantee', which, viewed through this lens, was an inevitable response to the ways in which customers were changing. However, viewed in another, it was entirely antithetical to the excitement of having a garden (in terms of waiting to see what will happen there).

Russell: And if you just have someone that wants to just get it sorted?

Lyn: You do exactly that. Um, I mean so much of it is container gardening these days, and you emphasise to them that you have to water it every day. But so many of them just don't and then it dries out and they come back and they say, look, this one went wrong! and that's tricky …

Russell: People take their plants back, then … as though they don't work?

Lyn: Yeah, its tricky, you know, if they come in and they say, well, we bought a plant off you and it died … They have to bring a plant back to me and then I will see. I won't just issue a refund.

But … It's Not Furniture!

Such was the exasperated claim of Izzy, a relatively new garden designer. She always seemed to me to be calm and kind-hearted – especially calm, according to her, after she made the move into this job. It's our first meeting together and we'd been talking through how she, and other garden designers, approached their work at the moment (and how they were likely to do so in future). One aspect was particularly interesting her: what to wear when visiting clients? In a recent meeting of the garden designer group to which she belonged they had been discussing this. For some, it really didn't matter – they were, after all, doing outdoor work and they were, as a consequence, expected to be 'practical people in practical clothes'. For others, however, they should indicate that they were providing an appropriately 'professional' service – and that meant dressing in same way as other professionals, namely suits and other 'smart' clothes. She hadn't thought about it before, but she thought that her recent dress choices suggested she was moving towards the second way of thinking – she had started wearing suits, so she wondered with me, because they seemed to fit with the idea of taking charge of her clients and underlining the idea that she could definitely give them what they wanted. Then the above exclamation followed (in which she suddenly railed against the idea of efficiently providing such a service).

But no-one ever said that it was like furniture. I certainly hadn't been suggesting this in our interactions just before. This was a view that she came to herself as, in the course of discussing these issues, she suddenly saw the implications of certain ways of working that could easily take over her. If designers treated garden plants like the 'furniture' found indoors, they might soon lose sight of all that was exciting about them in terms of watching to see what they would do in the garden! In that moment she realised how much she was against these ways of working and thinking (though she hadn't fully crystallised her position before this point). Another way related to the idea of 'waiting' for the garden to do certain things. Waiting could be difficult to discuss with certain clients when part of the excitement of hiring a designer stemmed from the suggestion of them leaving a relatively finished 'product' behind them. Viewed in this way, to talk of waiting for the garden to assume a particular shape, for certain plants to flower or for particular kinds of foliage to arrive was a problem as it immediately undercut that idea. To wait is also to be put in a subservient role. No-one (and especially not their 'central London' clients) wants to be made to wait.

So ... Do You Garden?

I'm in a client meeting with Anna, a recently established garden designer who works out of her studio in Brixton, South London. She was an estate agent before. But she wanted to do something that she 'loved' whilst still helping people with their homes. And garden design seemed perfect for that. The talk is flowing. She seems to have a good rapport with them. Then there is a pause. She stops and asks them 'do you garden?' I think that Anna would probably admit that this was probably the wrong strategy for the occasion – asking directly as though there should be a 'yes' or 'no' answer was not what many designers did. This was because if 'gardening' were introduced as a socially recognised, established activity (and if clients were generally taken to have little sense of how much they were capable of, or interested in, plant care), a positive response to the 'do you garden?' question was unlikely.

Instead, they often drew on other ways of talking about what might happen in the garden that made reference to other domains of life. As already discussed, many of the gardeners that I spoke with were adept at taking cues from what clients had around them at home. And this could bleed into how plants were spoken of. Many clients would, according to our discussions, talk in terms of 'low-maintenance' gardening. This was demoralising for many designers since to talk of 'low-maintenance' gardens was to see them as static. Or, perhaps more rightly, like a machine to be 'maintained' – held in place, not allowed to grow, to surprise or perhaps even to delight their owners. This related to a common client anxiety about 'staying on top of things'. In this thinking, the garden was 'built' and was thereafter 'done'. When I visited Danny, one of those who'd had their gardens relatively recently 'built' in line with a design, she was finding it hard to adjust to what she had been provided with – a variety of beds around a patio area and a further area for 'summer dining'. She had spent a great deal of money on making it a visually attractive place. And it certainly was, in her opinion, at the moment. But that was partly the problem in that she knew things could change. She liked the idea of having a garden, and potentially becoming 'a bit of a gardener', because that might help her find more 'balance in her life'. But, as things stood, what she actually had on her hands was a 'nagging worry'. There were still things that she 'needed' to do to make sure the garden was functioning as it should without her being required to intervene on a regular basis. She took some solace from the plan she had devised – 'it will be finished', she said, 'when the irrigation pump is eventually fitted'.

In this regard, the garden was unsettling to their ways of thinking about how homes should be organised in its potential for life, for growth and for change. It was unsettling to the practices of acquisition with which designers, their clients, garden centre managers and their customers were increasingly familiar (irrespective of how they'd ideally like their interactions to play out in principle). Evidence of 'personal style' was first identified and then translated into physical form. And it was often thought to be just too risky now not to work in this way, too difficult to coax clients into other kinds of relationship. Though, in effect, they wanted to 'talk them into their gardens', it was often easier not to.

Tentatively Encouraging the Enjoyment of Life

For Izzy, however, this was what a 'worn-out' garden designer would do. And she didn't want to provide all that hard landscaping, decking for dinning or 'corporate plants for corporate people' (plants that didn't do very much, and which faded into the background like those we might easily pass in the reception areas of some of the offices discussed in Chapter 3). She described that approach with a surprisingly resigned tone of voice for someone who had only been doing the job for three years (and was otherwise very happy doing it). Of all the ways in which she might work, this was definitely not what she wanted to do! This was certainly unlike how she had, moments earlier, been showing me a book of plants that she was excited about using in the future. This resignation was also rather out of character. This was because when she, and other designers, talked of being 'worn out' in this way, they meant no longer having the determination to attempt to share their excitement about plants (but rather running with the easiest way of handling clients).

Many of them had gone into garden design because they wanted to do something different to the 'usual rat race type of thing'. Yet still they found themselves falling into line with an idea that their job was efficiently to achieve design success. And that was done through the rapid production of a fixed garden 'product' that allowed their clients to socialise in certain ways outside or which conveyed a sense of style. This could make speaking of actively living plants (both with their clients and sometimes with me) a conversational challenge:

> 'And it's nice really that you are not that much in control of it. Plants, well, all of a sudden, they sometimes go and, of course, well, the clients want to know why has it died? And it's like, well, these things happen, I'm afraid!' (laughter)
>
> Margaret, interview two
>
> 'And I just sort of explain, and say well, you know, they are growing, living things and that's just the way it goes really, tough luck!' (laughter)
>
> Jenny, interview two
>
> 'Well, I can't guarantee that every plant will be performing at its best. It's in the right position to do well, but sometimes that's the way it works and plants that have performed excellently, don't do well at all.' (laughter)
>
> Izzy, observation session one
>
> 'Yes and, at the end of the day, plants are funny old things, and if they are happy, then you should just leave them there really, that's what I'd do. There are things which are slow and others that are there already ... and it will broadly sort of more of less get there in the end!' (laughter)
>
> Izzy, observation session five

If we think of their clients as shopping for gardens (and the job of the designer as one of providing relatively fixed products), we can see why laughter is needed to diffuse the tension of these situations. You wouldn't, for example, sell a piece of technical equipment and then say that, ultimately, it was impossible to know whether it would work for the customer. There was a sense here that, in intimating that plant behaviour could never fully be predicted, they were subverting the model of how things ought to proceed – talking out of line with the expectations for the situation, if you like. And that was part of the reason for the laughter through which they, sometimes nervously, acknowledged the essential reality of plants being ultimately uncontrollable. When Izzy deploys 'at the end of the day' she was, I think, alluding to how we are encouraged not to think of them as such. This related to how certain ideas about acquisition were, at the time, potentially enveloping plants, attempting to rebrand them as predictable products (even though they will never quite allow this 'at the end of the day'). But it was also about their personal enjoyment – about the difference that living plants made to the experience of working in and around gardens for them. In this way, their brief bursts of laughter also served to underline the genuine pleasure that they took from being subject to plant life – of being subordinate to the ultimately unknowable processes of the natural world, in effect. They liked them so much that they had to laugh.

That could then filter through into client interactions in ways that didn't always result in this pleasure being passed on. At a client meeting in East London, Grace is talking with a music executive and providing an overview of what could be done with space outside his home. She is walking around the current garden and she is animated. She is saying that this one plant is doing better than she expected in the circumstances (with a short laugh at the end). And she notes how another plant that she passes (pausing for a moment to mention the Latin name) is doing surprisingly well too. Hers is an upbeat stream of consciousness of a type to which I will return. It also (as she admitted herself afterwards) partly served to underline her expertise as someone with the required plant knowledge. Indeed, she was positioning herself as a type of person that designers sometimes talked of with a kind of hushed reverence – the 'plantsperson' (a knowledgeable and passionate horticulturalist who, in certain intriguing ways, somehow belonged to the plants). She was also showing herself as possessed of the kind of relationship with the garden that she wanted to encourage in her clients. And this was also one they often said they wanted for themselves too. It was one in which the individual was drawn into an ongoing relationship with plants that was transformative, making them into a calmer person, someone who took a less than fully determined approach to plant mastery. This was someone, in effect, who didn't think in terms of 'low' or 'high' maintenance, but rather in terms of being part of the process.

Yet, rather perversely, this could push the enjoyment of plant life away from the client. Though clients were taken to like the idea of assuming this kind of attitude themselves, being confronted with it could be jarring. In the case of the music producer, he had no idea about the Latin names (and no idea why the bush was doing 'surprisingly well'). And so it could sometimes be easier for the client – and ultimately for the designer too – to leave it to the designer to become the 'plantsperson'. Like the idea of doing 'gardening' (if framed as an activity with a specific kind of attached disposition), it could be tempting to speak as if assuming such a disposition was beyond them as an ambition. And so it could be easy to let the designer say they would do all the 'planting' (as a term that was sometimes used to gloss over the anxieties of talking about a diversity of potentially beguiling lifeforms that also needed to be kept under control if clients were to 'maintain' their new gardens). In truth (like Lyn who was tempted to 'palm them off with' the plants she already had in her centre), the designer sometimes encouraged this blanket treatment too because it meant they sourced the plants themselves (and that could make for a bigger profit).

In a Mad Prince Charles Way

Yet this is not to suggest that it was impossible for the designed garden owner to develop a more intimate relationship with the plants provided by their designers. Indeed, in some respects, features like the above laughter served as an invitation to join them in enjoying a sense of wonderment at how the natural processes that were channelled through their gardens could never be entirely controlled or fully understood. And, in truth, designers often hoped that this might happen. In that way, the above laughter was something more misty-eyed than an exercise in covering their backs in advance of clients potentially complaining about faulty plants. But speaking and thinking in this way could also be risky with clients who were often quite confident customers and who liked 'to get things done'.

Going back to Liz, who talked quite freely with me about relevant social trends and how others in the London were keen on 'getting into' gardening because that has come to represent an escape from modern life (and was also connected to how, as shown by the increasing numbers of tables she'd seen outside coffee shops, people 'felt Mediterranean' if they did everyday things outside these days). As with some of the office workers from Chapter 3, what was intriguing to me here was how she was positioning herself as apart from these wider populations even though she had evidently taken a strong interest in the design of her own garden (at least at the start). It was easier for her to talk about how 'modern society' wanted to reconnect with 'the earth' through these means than it was to admit that she potentially did too. And this was partly why some of this group ended up with a kind of 'garden machine' (in which plants did what they were told, when it was expected of them), rather than a place in which 'reconnections' happened. The challenge was partly about overcoming the difficulty of positioning themselves otherwise when they were more used to thinking and talking in terms of the efficient execution of tasks.

But this wasn't always the case. Polly was clear about the 'look' she wanted from her garden, for example, and about the activities that should ideally be accommodated within it (she wanted a space for her partner to practice Tai Chi and a dining area too). She had also worked hard on supplying the 'right brief' to her designer. But there were also some (often harder to articulate) ideas about making an 'urban oasis' that she expressed in a quieter, less confident voice to me as we sat in her kitchen looking out onto the lawn. She was worried, I think, about repeating a hackneyed phrase that she felt foolish about being associated with. We turned to how things had gone after the designer had disappeared and she was left on her own to find a way of living with the landscaping features and living creatures they had put outside her home. She became animated. 'And then it all started!' she exclaimed (like the plants of the garden had set off on a race and she was a spectator). 'And now it's bursting out … and I did quite a lot of standing and looking at things growing in a mad Prince Charles way – I just keep walking around and seeing what comes up.'

We will return to how she spoke of walking around and 'seeing what comes up'. But, for now, what is interesting in this exchange is the reference. Prince Charles is a member of the UK royal family who is sometimes pilloried by the national press for apparently eccentric ideas about how people should relate to the natural environment. Though a keen champion for the idea of living sustainably, he's sometimes decried as out of touch with modern life, as 'away with the fairies', because of a perceived naïve hopefulness about how we could salvage traditional lifestyles, and how people might be happier as a result. In the above section, Polly is making fun of herself. She is recognising that there is a certain comedy associated with relating to the natural world in this way (or, at least, that she too might be ridiculed if she thought too much in this way). She was like Liz, who preferred to evaluate others when talk turned to the abstract pleasure of living with plants (showing she was able to stand apart from them and offer a sober analysis in our interview exchange, rather than reporting on her own experience of jumping into it and enjoying it too). Polly was a successful businesswoman who worked in marketing. And the idea of simply 'standing there and looking' was not something that seemed to fit with how she tended to speak about her actions (and saw herself as a consequence). Yet, though often quietly and cautiously, she described such experiences in our interviews, nonetheless. The actively living garden had managed to trigger an alternative way of acting in, and talking about, it.

In this respect, and in a similar way to how the office workers spoke of enjoying what was happening 'outside' – the idea of 'play' was evoked by some of the designed garden owners who had 'got into it' (with the implication again being that 'nature' was something for children really, that they were generally too busy for that now). The designers were often keen on framing it as such too (as a means of making the activity less pressurised and intimidating than the more clear-cut idea of effective 'gardening'). They spoke of giving clients a small 'area to play with' – a 'safe space' in which they could test out the extent to which they wanted to get involved without the stylishly designed environment they had been paid to supply being endangered. But, unlike Stefan in Chapter 3 (who ironically suggested that he and his work friends liked to 'go for a wander'), if all went well, these garden owners were actually doing it. So, by learning to talk in this way (and learning to position themselves as less than fully determined busy 'modern' people), they took some of the pressure off themselves. And that meant they were permitted to become less than appropriately focused people (the kind, of course, who often end up being able to afford a garden designer!). So, these were sometimes tentative and often quite self-aware forays into thinking and talking about their garden relationships differently – and doing so in ways that they often wanted for themselves (but currently found challenging to assume).

Compare the awkwardness that followed when Anna asked, 'do you garden?' to the below interview extract in which Kirsten is discussing what she did with her garden:

'We won't learn it because we are not keen gardeners; we just do the things that we know, like weeding. In the first few years we tried to learn it, how to do the roses and when to cut ... we dared to touch the plants and we learned to do that. But I knew that I will never learn this, and in this country, they have the greatest gardeners in the world ... so why should I end up knowing it?'

Rather than being an exhibition of modesty, rejecting the title of 'gardener' can perversely allow you to get closer to the garden, as the expectations associated with this moniker are artfully sidestepped. Effectively Kirsten is doing the same as 'having a little area to play with' in the sense of positioning her actions as much less determined. Through saying that they are definitely not 'gardeners', they put themselves in a better position to relinquish control in a way that allows them to, at times at least, become more involved with the life of the garden (which, we could argue, is actually what being a real 'gardener' is all about).

A Different Sort of Silence

Sandrine would probably be considered a success story amongst the garden de-signers. She was what they wanted or, at least, what she became was what they hoped would happen to their clients after they left. For the most part, though they talked as if a big part of their job was to leave happy clients with something that was immediately visually appealing (and though they knew that those for whom they worked were well versed in thinking about the garden as a backdrop for various social activities), they ideally wanted to leave them with something different. Essentially, they wanted to ignite an interest in being part of the on-going process of living greenery, which was why they personally loved working with gardens. Some even felt they had a degree of responsibility to make that happen. Izzy, for instance, 'took a very dim view' of the idea of talking too much with clients about 'outdoor rooms' even though she felt this was an idea that it was 'safest' if she partly accommodated. Forthright Margaret took a pragmatic approach to this challenge by, in effect, teaching her clients a different way of relating to the garden such that by the end of the process they were 'using the same vocabulary' as her. Grace, to link back to the troublesome idea of 'wait-ing' for the garden, deliberately spoke of garden processes 'taking time'. Her thinking was that doing so would 'jolt them out' of their usual way of getting things quickly organised that would (if she didn't intervene) bleed into how they handled their gardens too. By emphasising how plants gradually grew, she was forcing them to acknowledge that their gardens contained an ongoing process of growth, development and decline – an ongoing process they might potentially get involved in themselves.

Now back to Sandrine. At the start of the design process, so she excitedly told me (over wine and some nibbles offered on a small plate inside her ground floor flat), she thought that she would end up 'sitting there and just looking at things' – sitting and looking at the garden just as we were, at the time, doing ourselves. But then, she recalled, she 'started to move things around'. Telling here, I think, was her use of the passive voice (she found herself doing something – this was far from the determined execution of a plan). She also said that things 'just started happening' in her garden. She was no longer in charge of a fixed garden 'product'. She was part of something else. And she enjoyed that. When we took a short tour of the garden after our initial discussion, she fell into silence. But this was a very different sort of silence to that of Liz in the second example with which I began my analysis in this chapter. Unlike Liz, she was not indifferent to, or lacking in things to say about, the plants. Rather she has been absorbed by the garden. She was 'just doing things' as 'they occurred to her' in a way that was transporting and of which she was only partly aware.

Conclusion

Talking Customers into It

Were I, as I did for the office workers and the treadmill runners, to attempt to crystallise the overall process that I observed in my domestic garden fieldwork into a single summative verb, I would say that garden owners were (in some cases, and if they were lucky) eventually succumbing to the outdoors. This was a case of gradually letting go of the impulse to control the materials that came into their gardens, to get thing efficiently and effectively organised so as to create immediate 'impact' through the determined execution of some design or another (as though they were importing shopping registers directly into their gardens). And, for those who eventually let go, the outdoors literally became an absorbing place in the sense that they became part of, instead of separate from, the ongoing processes which garden life comprised. In that regard, the garden eventually coaxed them out of the imperative to act in line with a certain kind of shopping practice.

Succumbing also fits well, I think, because it hints at the challenges involved (and how the process can happen slowly) when, as the concepts on which I am drawing would have it, the practice at hand, namely in this case shopping, doesn't always release you as easily as all that. Paying attention to the challenges involved helped to explain a strange and frustrating situation for some – how they could be drawn to the idea of spending time in the garden because they hoped this would be relaxing but ended up buying a garden that didn't particularly need them there. This was a difficult situation for some of the garden designers to negotiate in their interactions too, especially when they were personally alive to the pleasures of growing. Getting their clients to develop such a relationship required some work on their parts as they tried to coax them through a third environmental unsettling.

In terms of understanding how it is that, for example, expanses of hard landscaping may be crowding out plants, grass and trees in the domestic gardens of London (and recognising how other factors are involved too), this project shone a light on how certain interactions play a part. By organising my thinking around how the practice of shopping might be unsettled by the involvement of living plants, I was able to notice the challenges of learning to relate to gardens otherwise. Returning now to my broader ambition, the question that follows is what does this mean for beneficial garden greenspace experience? On the one hand, the implication is that people who shop for their gardens and who end up not engaging that much with them by making them more firmly 'material culture', might end up with less of the benefit than they might have otherwise hoped for. On the other hand, however, it is quite possible that gardens could still be full of greenery, just greenery that requires comparatively little care. In that sense, we might ask what the problem is here. Though such gardens might, for example, not be so good for wildlife they could still provide a beneficial experience for owners who look upon them. However, the challenge that this project pointed to in this respect was about whether those who are more habituated into getting things quickly and effectively organised are inclined to stop and look. In that sense, one of the conclusions was that garden designers might continue to resist being 'worn down' by the ways in which a certain kind of shopping practice was seeping into their interactions and shaping the implied ideas about what a garden should be. In other words, the idea was to remember that they could sometimes 'talk their clients into' their gardens in ways that might benefit them. Now, of course, not all garden owners are in the enviable position of being able to enlist the help of a designer. But, by starting with a wider set of dispositions linked to shopping as a commonplace practice, I saw the difficulties of growing gardens when it can often feel more familiar to shop for them instead.

Making Light of Something Serious

In this study, along with talking with some of those who were sometimes coaxed out of a shopping practice, I also watched attempts at doing so and spoke with some of those who made these attempts in my interviews. This allowed me to consider my three suggestions from a new vantage point. And, though there were similarities, shopping for a garden was still quite different to working in a corporate office or regularly running on a treadmill.

- To return again to the first (about the talk produced by practices), in the last chapter, I noted how people were aware of how practices can create particular forms of talk and how that can encourage some of them to 'have all the chat'. Here we saw how certain formats of 'chat' can be easy to fall into because of a degree of familiarity with the practice that encourages them. These were the kinds of interactions that the more 'worn down' garden designer might stage – efficiently rattling through a series of questions that were more easily handled by those who had done plenty of shopping but not too much gardening. The idea of gardening, by contrast, didn't call forth forms of talk so readily. And that was, without doubt, partly related to how I was speaking with certain groups of people. Yet, it was also about how gardening was often a solitary practice in which the mental conversation might be as much with plants as people. And that was part of the trouble. When shopping came with certain familiar patterns of speech, plant conversations could suffer by comparison.

- In terms of how people managed the experience of speaking about the practice at hand, the most notable example related to the contextual deployment of laughter. Though perhaps not always consciously done by the designers who embarked on their nervous laughs, they were attempting to diffuse the situation when acting out of step with the shopping practice. And those who came to acquire, and enjoy, certain kinds of garden could also draw on humour to make light of how they had effectively succumbed to the associated pleasures. When Polly joked about the 'mad Prince Charles way' in which she had started to relate to her plants, the point was that speaking in this way also helped her to keep going. By acknowledging how she was acting out of step with how she, and some set of imagined peers, tended to act more widely, she could more easily get on with the enjoyments of adopting an alternative approach. Conversations were being managed here (though in only a semi-conscious way with reference to the idea that they ought to be focused and purposeful people).

- The assumption embedded in the rationale for making light of this situation is that most other people probably don't think and act in that way. And that was where I saw evidence for the third suggestion, namely that, under the circumstances of being encouraged to reflect upon otherwise unthinking actions, people often reach for ideas about what others would do. Beyond that, it was harder in this case to discern too much evidence for this. This was, in part, because the practice that I wanted to start with, namely that of shopping, was still in the process of seeping in the gardening context (whilst also being resisted there). In other words, the idea of shopping for a garden isn't so well established that ideas about how others were doing it were readily available. But that was not to say that connections could not be made. And, indeed, talking could encourage people to see how the practice was creeping into this context. When Izzy exclaimed 'it's not furniture' she had snapped into a sudden awareness of what others were potentially doing. And as the idea of the 'central

London client' spread amongst the designers and the managers, notions about developing norms of action were being identified and exploited.

Unlike in the office where the focus is on 'work', and on the treadmill where the idea of 'working out' was part of what kept people running where they did, we might have expected the garden to be different. Yet, despite the suggestion of many Britons 'loving their gardens' that was introduced at the start, we have now seen some of the challenges of getting to the point where they could 'love' them as part of an ongoing connection. Getting to that point was also linked to a developing picture of how many of those that I spoke with in my studies felt the need to be purposeful. There now remains a single case to consider. I've left it to the end because perhaps it is here that alternative states are most fully assumed – after all, isn't that part of the whole purpose of getting away from the city?

References

Baker, L., Boyer, C., and Hikaru, H. (2018). Online opportunities: a quantitative content analysis benchmark study of online retail plant sales. *HortTechnology* 28: 516–523.

Bhardwaj, V. and Fairhurst, A. (2010). Fast fashion: response to changes in the fashion industry. *The International Review of Retail, Distribution and Consumer Research* 20: 165–173.

Bhatti, M. (2006). 'When I'm in the garden I can create my own paradise': homes and gardens in later life. *The Sociological Review* 54: 318–341.

Cameron, R., Blanusa, T., Taylor, J. et al. (2012). The domestic garden – its contribution to urban green infrastructure. *Urban Forestry and Urban Greening* 11: 129–137.

Cass, N. and Shove, E. (2017). *The last mile and the next day: the changing times and spaces of shopping – implications for energy demand*. Lancaster University working paper.

Freeman, C., Dickinson, K., Porter, S. et al. (2012). 'My garden is an expression of me': exploring householders' relationships with their gardens. *Journal of Environmental Psychology* 32: 135–143.

Gatersleben, B., Jackson, T., Meadows, E. et al. (2018). Leisure, materialism, well-being and the environment. *European Review of Applied Psychology* 68: 131–139.

Ginn, F. (2017). *Domestic Wild: Memory, Nature and Gardening in Suburbia*. Abingdon: Routledge.

Global Data. (2018). *Gardening*. London: Author.

Gross, H. and Lane, N. (2007). Landscapes of the lifespan: exploring accounts of own gardens and gardening. *Journal of Environmental Psychology* 27: 225–241.

Head, L. and Muir, P. (2007). *Backyard: Nature and Culture in Suburban Australia*. Wollongong: Wollongong University Press.

IBIS. (2019a). *Garden Centres and Pet Shops*. London: IBIS.

IBIS. (2019b). *Flower and Plant Growing in the UK*. London: IBIS.

Masnaghetti, M. (2016). When does spring start? Trends in the gardening category. *Journal of Direct, Data and Digital Marketing Practice* 17: 236–243.

MINTEL. (2018a). *Garden Product Retailing*. London: MINTEL.

MINTEL. (2018b). *Clothing Retail*. London: MINTEL.

Ottosson, J. and Grahn, P. (2005). A comparison of leisure time spent in a garden with leisure time spent indoors: on measures of restoration in residents in geriatric care. *Landscape Research* 30: 23–55.

Parr, H. (2007). Mental health, nature work, and social inclusion. *Environment and Planning D: Society and Space* 25: 537–561.

Pitt, H. (2014). Therapeutic experiences of community gardens: putting flow in its place. *Health and Place* 27: 84–91.

Pretty, J. (2004). How nature contributes to mental and physical health. *Spirituality and Health International* 5: 68–78.

Scott, T., Masser, B., and Pachana, N. (2015). Exploring the health and wellbeing benefits of gardening for older adults. *Ageing and Society* 35: 2176–2200.

Smith, C. (2010). London: garden city? Investigating the changing anatomy of London's private gardens, and the scale of their loss. London Wildlife Trust Greenspace Information for Greater London Greater London Authority.

Söderback, I., Söderström, M., and Schälander, E. (2004). Horticultural therapy: the 'healing garden' and gardening in rehabilitation measures at Danderyd hospital rehabilitation clinic, Sweden. *Pediatric Rehabilitation* 7: 245–260.

Soga, M., Gaston, K., and Yamaura, Y. (2017). Gardening is beneficial for health: a meta-analysis. *Preventive Medicine Reports* 5: 92–99.

Taylor, L. (2002). From ways of life to lifestyle: the 'ordinari-ization' of British gardening lifestyle television. *European Journal of Communication* 17: 479–493.

Taylor, L. (2008). *A Taste for Gardening: Classed and Gendered Practices*. London: Routledge.

Thompson, R. (2018). Gardening for health: a regular dose of gardening. *Clinical Medicine* 18: 201–205.

Timms, P. (2006). *Australia's Quarter Acre: The Story of the Ordinary Suburban Garden*. Melbourne: The Miegunyah Press.

Uglow, J. (2017). *A Little History of British Gardening*. London: Chatto and Windus.

Zukin, S. (2005). *Point of Purchase: How Shopping Changed American Culture*. New York: Taylor & Francis.

Chapter Six
Embracing the Outdoors: At the Festival

The Intriguing Challenge of 'Outdoor Hospitality'

Though holidays involving basic sleeping structures in environments that are often both green and far away from our cities are established leisure activities, they
have recently enjoyed a resurgence of interest in the United Kingdom (MINTEL
2018a). This is no doubt partly because they provide a comparatively cheap
means of escaping the usual experience back home (O'Neill, Riscinto-Kozub,
and Van Hyfte 2010). But it is also, according to some analysts, about the particular pleasures that come from participation. Some have even argued that the
resurgence of camping is a response to some of the pressures of modern city life
that were seen in the three other case studies presented in this book – pressures
about appropriate self-presentation such as those felt by the office workers, pressures of perceived lack of time such as those which made some of the treadmill
runners stay firmly where they were, and pressures to do with being sufficiently
efficient in responding to tasks such as those embodied by some of those who
turned to the garden designer. Some have said that camping trips, for example,
are attractive precisely because they take people away from these pressures and,
through these means, somehow help them to 'reconnect' with the natural environment (Brooker and Joppe 2013).

But how easy is it to achieve these 'reconnections', and what should they ideally involve? These are more than merely academic questions for an increasing

The Unsettling Outdoors: Environmental Estrangement in Everyday Life, First Edition. Russell Hitchings.
© 2021 Royal Geographical Society (with the Institute of British Geographers). Published 2021 by John Wiley & Sons Ltd.

number of businesses for whom this resurgence features as a tempting source of profit. And, in doing so, they are thinking hard about how exactly they should 'cater' to a potentially expanding market of campers when it seems that what many of them want to escape is the very idea of being catered to (better to 're-connect'?). Certainly, for the 'spartan campers' (Bultena and Klessig 1969) – an imagined group for whom self-identity is linked to the personal improvement that comes from outdoor challenges – this is the case. Meanwhile, for those who we might see as their less adventurous 'convenience camper' cousins, the draw of familiar facilities might be much stronger. Either way, the right way of furnishing campers with the most desired levels of 'outdoor hospitality' is not immediately obvious (Brooker and Joppe 2013). What kinds of 'customer service' should be provided to people when part of the attraction may be about escaping the idea and implications of being a 'customer' in the first place?

Some have generated evidence for those businesses that walk this definitional tightrope by studying the changing ways in which campers speak about what the experience gives them (Garst, Williams, and Roggenbuck 2009). And, whilst some of their reported motivations have stayed relatively fixed over recent decades, others seem to be in the ascendant. For example, whilst campers continue to tell themselves conflicting stories about the co-existing appeals of escape and convenience, there seems to be a growing tendency to talk in positive terms about visually experiencing the natural world – of being able to 'contemplate nature' without having to engage with all the dirt, dust and discomfort that sometimes comes along with it. Perhaps people increasingly like to look at, and then re-cord, natural scenes (Cordell 2008) as social media and the increasing ubiquity of digital cameras and smartphones continue to reshape societies. This takes us quite nicely to the cultural advent of 'glamping' (see, for example, Brochado and Pereira 2017, for a recent consideration) as a term that combines 'glamorous' and 'camping' and features as a holy grail of profit for a growing number of businesses. Their hope is to find the perfect combination of facilities and aesthetics that will tempt a new set of consumers to experiment with countryside accommodation that dips slightly below the standards with which they are more familiar. They do so by providing some services and not others, perhaps by erecting tents or yurts that holiday makers don't need to go to the trouble of putting up themselves (and which also look good in their photos).

This raises questions about what camping is doing for those involved, and whether it is continuing to provide some of the benefits with which it has long been associated. Camping is commonly linked to 'wilderness therapy' (rather like the 'horticultural therapy' of Chapter 5) – the suggestion that spending a period of days living a comparatively simple life in natural environments can trigger a whole host of psychological and social benefits (Russell 2001; Pretty 2004). However, unsurprisingly enough (and again like the studies of therapeutic horticulture), the work that we have on this topic tends to focus on particular groups who we may particularly want to derive these benefits. Children and adolescents who many have worried about suffering from increasing amounts of 'nature deficit disorder' (Louv 2005) provide the most common focus. Studies have tested the suggestion that camping experiences might make them both more mentally relaxed in the present and potentially more committed to the environmental cause in the future (Wells and Lekies 2006; Collado, Staats, and Corraliza 2013). There is also work on soldiers recovering from traumatic experiences and those already facing various mental health challenges (see Morris 2003, for an overview).

Despite the evident value of this work, by focusing on those who particularly warrant our attention and assistance in the present, it doesn't say so much about the changing ways in which wider populations relate to these experiences. In other words, it tells us less about the broader social trends that are having an influence on all of us. In this respect, one issue that might feasibly be putting many people off is about the immediate physicality of the camping experience: how comparatively simple modes of outdoor living might effectively unsettle certain ways of managing our bodies to which we have somehow become wedded. It is with this in mind that I now turn to personal washing. In questioning the advent of 'glamping', MINTEL (2018a) suggests that young people are comparatively uninterested. This is partly because they cannot afford it. But it is also intriguingly because many of the young people who, in difficult economic times, have ended up experiencing the 'forced dependency' of living for longer at home with their parents may be especially interested in testing their 'survival skills' through something more 'spartan'. Though this may be true, they also, as we will now see, seem to be enjoying certain creature comforts more than most.

The Societal Spread of Showering

As with the preceding three chapters, my fourth empirical chapter now turns to a practice that is widespread and for which the changing ways in we do it could have implications for the likelihood of beneficial greenspace experience infiltrating people's lives. The other three case studies can be relatively easily connected to greenspace were we to reflect on them (though, as we have seen, this may be more easily said than done). We may like to think of office work being punctuated by some time outside the floating factories that dominate many city skylines. Running is still commonly, at least imaginatively, connected to outdoor parks. And, even if we are increasingly shopping for our gardens, for the moment many of the things that we buy are still likely to be living plants. My final practice requires a bit more of a mental leap to see the connection at the start. But I think it is one worth taking. The practice that provides the focus now is showering – the idea of washing our bodies by standing under a water source and allowing that water to stream over us as we endeavour to clean ourselves whilst also, and increasingly, enjoying the experience of lingering there.

Current washing habits are very different from what they once were (Hand, Shove, and Southerton 2005). The ancient Greeks, for example, enjoyed collective baths, partly because the technologies of the day were suited to this and partly because of certain beliefs about the civic benefits of bathing together with others. Indeed, it can be difficult to disentangle the mechanisms that have shaped how ways of washing have changed over time (though some researchers interested in encouraging societies to take less resource consumptive paths have tried). For example, the journey to where many societies currently are with their washing can be seen as a collective outcome of changing ideas about health and the value of cleanliness: how washing has gone from being associated with ideas about effective dirt removal, personal disorder and potential disease to current preoccupations with being appropriately fresh and fit for public consumption. Another way of looking at it would be to focus on the increasingly private nature of water provision: how predictable water supply was linked to the development of domestic shower technologies and bathroom cubicles. A third way would be to turn to the temporal organisation of everyday life: how people collectively feel that they should organise their time and how weekly baths were replaced by speedy showers that chimed better with lifestyles that, at least, felt increasingly rushed.

There are many contributors to the societal spread of showering. But what is definitely clear is that it has spread. Certainly, Britons have been showering more often (and for longer) than ever before (Pullinger et al. 2013). Indeed, showering once a day has, for very many people, become 'normal'. However, variations remain. In Denmark, for example, we know that some Danes tend to do so in the morning and others like to shower at night (Gram-Hanssen et al. 2019). This depends on a range of factors but includes how, to link back to Chapter 3, those who work in offices feel obliged to present themselves in certain ways there (whereas others who feel less pressure to appear 'fresh' may prefer a relaxing evening wash instead). Then, in the United Kingdom, we still see the co-existence of various strategies within the same society (though it does seem that the bath is on its way out and the days of 'flannel washing' are also numbered) (Pullinger et al. 2013). Viewed in this way, the normalisation of significant everyday showering is troubling because of how it may be 'entrenching and establishing new benchmarks for smell, hygiene and presentability' (Strengers 2009, p. 8) that entail significant resource consumption. This is partly supported by evidence that suggests that those who wash less often (and who do so in the bath) are those who grew up in times when doing so was more widespread. Younger people, by contrast, have, from the start, been exposed to a society typified by frequent washing to the extent that this could, in fact, now (rather perversely) even be detrimental to public health (Browne, Jack, and Hitchings 2019).

This raises questions about practice transmission – how it is that our young people have learnt to wash so much? This question is especially intriguing in this context because, compared to the three practices that have informed my analysis so far – working, running and shopping – it is most often done (and increasingly done, if we think of the decline of communal facilities) in private. In view of this, how do they learn to conform (if that is the right way of understanding the increasingly frequent showering of young people)? Some have suggested that peers can be a particularly strong influence. Though children may initially emulate their parents and be naïvely unaware of wider expectations and anxieties, during adolescence and afterwards the pressure to follow your generational peers can start to kick in. Gram-Hanssen (2007), for example, offers the example of one young respondent in her study who suddenly realises that their washing actions are out of step with their classmates. And then, in their cleanliness 'coming of age' moment, things suddenly shift and then don't go back. Waitt (2014), with reference to the hot and sticky summers of South East Australia, observed similar anxieties when he found that it was often only in the home that his young female subjects allowed their bodies to embark on the natural physiological response of sweating. Elsewhere, the fear of public condemnation could prove too strong.

These accounts encourage us to consider the processes through which people, particularly young people, learn from one another. As part of that process, one aspect that I particularly consider in this chapter is how they speak about a private practice brimming with potential social judgement. But, for now, the crucial points are that showering is more common than ever before in the United Kingdom and, whether they think about it or not, this seems to be especially so for younger people. In this context, it is interesting to consider how groups of young people might collectively cope with disruptions to such a situation. More specifically, and looping back to the earlier camping discussion, how might they handle an experience of outdoor living that might require them to question their attachment to the shower?

And that is the final way in which the outdoors is taken to unsettle an otherwise widespread social practice in this book. It does so by turning to a kind of naturally occurring 'experiment' in alternative ways of living with the outdoors (Browne, Jack, and Hitchings 2019). In doing so, we now travel to sites that are 'green' (in that they take place in fields with trees around). But the idea here is not to focus too fully on that (when these features are seldom a central draw for those involved). Instead, the aim is to use this site to consider how welcome the disruption of comparatively simple living is amongst the young people who predominate there. In that sense (and in line with my essential ambition for the book), the idea again is to take a few steps back from the greenspace focus, to understand better how easy it is to be in greenspaces (when, if we went, certain social practices are unsettled in the process).

The Summer Music Festival

Enter the summer music festival. Such events have been an increasingly common feature of the European calendar (MINTEL 2018b) as a growing number of attendees (most often young people) spend a few days camping in fields. Though the camping experience may feasibly be part of the draw, their main objective is to enjoy the performances and spend some time relaxing with friends. In this sense, the summer music festival can be seen as a context in which those who go there are forced to reconsider otherwise entrenched ways of personal washing, to calibrate their actions against those of others in a comparatively public setting and (in a few cases) to talk to a team of researchers about how they handle all that.

Some existing festival studies hint at how this can work for festival goers. One of the most common features explored in this research is how the enjoyment of attendance partly stems from how festivals present an 'escape' from, or 'alternative' to, the apparent drudgery of everyday life back home. In historical terms, after all, we know that festivals have long served as an effective means of maintaining the status quo by providing a kind of societal 'safety valve' (Anderton 2008) as revellers are allowed temporarily to subvert wider norms before getting back to their work (Getz 2010). Partly because they often take place in rural locations, festivals attendees are commonly taken to relish the opportunity to live differently to how they do back in their cities. As such, Browne (2009) discusses how a women-only festival in the United States can be exciting but is also something for which potential attendees must be 'ready'. Jaimangal-Jones, Pritchard, and Morgan (2010) describe how the appeal of the summer music festival, in their understanding, stems from entering a 'liminal' space in which unusual experiences are expected. O'Rourke, Irwin, and Straker (2011) similarly see festivals as sites that are explicitly apart from the pressures of everyday existence as rural locations help attendees to shake off the constraints of appropriate urban civility (see also Flinn and Frew 2014). In this way, summer music festivals can be linked to a nostalgia for 'simpler' living (McKay 1994). Lea (2006) details how this can sometimes happen through a gradual 'unwinding' process that can result in new relationships between attendees and their bodies. Finally (though we don't know for how long this is imagined to last), Purdue et al. (1997) think of the music festival as a kind of 'cultural laboratory' in which predominant societal norms are temporarily scrambled in the shaping of experimental new identities.

What this all means for personal washing is currently unclear. Perhaps surprisingly when the outdoor festival can be both celebrated and vilified in the media when attendees revel in muddy conditions after rainfall (Anderton 2008), how they more generally handle an environment that could easily be labelled 'dirty' has been only tangentially addressed in terms of its potential role in the apparent 'escape' of the festival. Though some of those in Browne's (2009) study talk about communal showers representing an enjoyable return to 'the land', she found others determinedly downplaying these conditions when enticing potential recruits. Similarly, for regular attendees in New Zealand, there is a fine line between mud adding to 'the memories' and the remembered disgust of 'walking in filth' (O'Rourke, Irwin, and Straker 2011). In Denmark, Andersen (2013) notes a similar kind of contradiction. His study highlights the enjoyable communal experience of queuing together for showers. But he also points to how the festival 'liberates' people from 'norms of etiquette and decency' (that are presumably partly why they are queuing in the first place). Recent UK market research (UK Festivals Awards and Conference 2018) also suggests a degree of attendee ambivalence about their festival washing practices: 23% said one of the things that they most loved about being at the festival was the 'sense of escapism'. Yet 7% also said that 'roughing it without showers or clean toilets' was one of their 'biggest downers'. The questions that naturally follow from this relate to whether everyday washing practices find ways to live on at the festival and how easy it is for attendees to escape their grip.

Getting Stuck In

My fourth and final case study focuses on how young people at UK festivals negotiated this ambivalence together, and how close attention to how they spoke of their personal washing before, during and after the festival illuminated this process. In this respect, this chapter turns away from an examination of how talk can help us understand how everyday lives generally go on in society to consider what particular speech patterns tell us about perhaps the most remarkable outdoor unsettling of the four – being faced with a very different way of washing yourself when living alongside others in fields of tents. Eager to explore how this process would be narrated to us, together with three other researchers, I travelled to two UK festivals in the summer of 2013. Both took place away from the city in fields that were surrounded by trees. In one of them, an area of woodland was even incorporated into the festival as a place for escape. In that sense, both took place in greenspace. However, the green of some fields was soon almost entirely eclipsed by a kaleidoscope of coloured tents and the grass was often soon worn away. Also, depending on how you look at it, we were lucky in that the weather was mild for both, ranging between 10 and 22°C and with little rain. This made for a comparatively easy camping experience. Ours was therefore a study of sweat and sun and summer living outside more than one of how it was to sleep in fields that had become dauntingly unattractive expanses of mud as a result of how recent rainfall was compounded by the effects of many people walking across them.

We settled into a shared tent with our Dictaphones and some conspiratorial exchanges about how our fieldwork should proceed. We decided that it was very probable (and indeed this proved to be the case for most of those with whom we spoke) that the majority of festival goers were unlikely to shower during their time there (both festivals lasted three to four days). We also figured that we would probably do better interviews if we could relate to them more fully in this respect. And so, with an interesting, and sometimes difficult to articulate, nervousness about how this would be, we resolved not to shower whilst we were there either. Beyond that, our basic objective was to learn about the cultural response of young people to being outside and away from the showers under which they seemingly spend a significant amount of time at home. Our own experience of shunning the shower, we hoped, would make for some rich pickings in terms of discussion topics and strategies for grounding our interview talk in the lived reality of the festival experience. We split up and approached people. In total, we spoke with 60 attendees aged under 40 for around 30 minutes.

We often sat with them on the grass, at benches around the festival site or even on fallen tree trunks in woodland areas. Once settled, we'd embark upon some seemingly casual conversations in which we explored the issue of what the outdoor festival experience does to otherwise entrenched ways of washing. The individuals we approached were often waiting for friends and so the interview sometimes morphed into a group discussion when the others eventually turned up. This sometimes proved particularly revealing in terms of showing us how groups worked together to establish a shared perspective on the topic and how individuals spoke of an otherwise private practice in front of their peers. On occasion, our respondents turned the tables on us. And that seemed to fit with how, at the festival, everyone was an equal. This involved grilling us on what we were doing before interjecting with observations and thoughts that they thought might be relevant. Often (and again in line with cultural norms at the festival) humour found its way into our conversations. And that was, of course, of interest to us too. Certainly, as was the case with all four of this book's studies, when this happened, this was seen as an opportunity for analysis rather than an obstacle in the quest for the truth of their perspective: how individuals and collectives reacted to our questions was as revealing as any of the concrete answers that they gave.

That Never Happens

I'm sat in an elevated festival field on the Isle of Wight. This is a nice interview spot, I think to myself, overlooking the whole site. It almost seemed to invite reflections on what the festivals was (and could be) as I watched its workings from afar. The time is around 11.00 a.m. and another sunny day is predicted. People seemed excited and I get several 'hellos' from passing strangers. This is a good time for an interview too, I thought. The attendees who could be seen down in the main festival site below were groggily waking up. They were peering out from their tent flaps and setting off in search of breakfast. They were probably mulling over last night's events together, I figured. I'm with Jenny, a veteran of several festivals, and a young professional from London. As with many of our interviews, I began by getting some context on how bodily washing usually went on for her back home.

She, like many of the respondents in this fourth study, straightforwardly said that she 'showers every day'. However, further questions revealed that she didn't always. 'I shower every day' was rather a useful turn of phrase – both for her and for others in the study. It was a resource on which to draw in the moment, partly because, in this instance at least, it allowed them to present themselves in a particular light (as though they were following a positive routine, as though they had internalised, and were living in line with, some sort of vague public health message). But this response also proved handy (more interestingly for my purposes here) because it effectively stopped the person from needing to engage in any further thought on this matter – in saying this, they were drawing a line under the matter with an apparently categorical statement that pushed further examination off the table.

Though such a response (in the moment, at least) did not invite further enquiry, part of what we hoped to explore was how it was to have (or perhaps belong to) this kind of routine washing relationship. So we persevered with this topic even though holding our nerve demanded some confidence when we were expected to move on. I asked Jenny whether she ever did anything different. In so doing, I was partly emboldened by how the festival context itself naturally seemed to invite an honest and open examination of some otherwise fairly fixed social practices, even with some nosey academic strangers such as us. Her bald response to me was 'that never happened'. And so, for someone who was otherwise quite intrigued by norms of bodily management at the festival (elsewhere, as I'll discuss, Jenny had a lot to say about when and where it was appropriate for people to pee at the festival), here I saw a passive voice taking over. It 'never happened' not because she didn't want it to, but because repeated enactment meant she no longer thought about it. Active choice didn't come into it – some things just happened, and some never did.

Others experienced similar difficulties. A number of the festival goers found it much easier to talk through the specifics of their routine (this is when and how I wash) than to speak of the logics that might be underpinning these actions (what are we trying to achieve when we shower?). They knew the procedure in some detail, but they had long forgotten (if they ever really knew it, or if there ever really was one) the rationale. Showers, as one respondent put it, were 'just nice, nothing more'. In other words, frequent showering was 'just what they did'. As another suggested, maybe it was just a 'psychological thing'. This was a term that repeatedly popped up in all four of my studies when respondents made appeals to mental processes that were important but also somehow unavailable to them there and then. Perhaps it was something to do with 'being awake' in the morning more than necessarily 'being clean'? Prudence said, after a lengthy pause (in an interview that she 'greatly enjoyed' as a means of understanding her 'ways'), that morning showering, for her, was 'a trigger for legitimate and constructive activity'. Though she did various things at home beforehand that could feasibly be construed as 'work' (emails in bed, making plans over breakfast), it was the shower that signalled the start of the legitimate 'day proper' – it marked a transition to a determinedly productive engagement with the wider world.

Partly because the festival was itself, as I have already mentioned, a relatively 'reflexive' space (a place apart from more usual patterns of living that helped us in examining them), the festival context itself helped us to explore relationships with everyday showering. And, through this exercise, we saw how there were few readily available answers to the 'whys' of their usual showering practice back in the city (in other words, when we asked respondents to supply underpinning rationales, they were rarely there). This was much more a story of how some things happened because 'they just did'. Going back to the bald 'that never happens' statement from Jenny, a comparable response to the question was provided by David, who categorically stated he was 'not willing to entertain the thought of washing less than once a day'. Part of the reason why he wasn't, of course, was because any thoughts of doing otherwise had gradually disappeared from his life – he wasn't willing to 'entertain the thought' because long ago such thoughts had been put aside after (hopefully) setting up the right showering practice for him. And so 'entertaining' them now was just too much of an effort in terms of digging them up, dusting them off and evaluating their worth.

I Like to Think That I'm Fairly Hygienic

Though one of the most commonly reached for justifications for frequent showering was about 'being clean' (and though, as just discussed, this seemed more like a rehearsal of the 'acceptable response' than anything else), 'being clean' didn't often figure as a particularly cherished state. Most often it was connected to the idea of escaping social condemnation. Having a sense that you were personally doing what was collectively deemed appropriate was paramount here. Certainly, hopeful appeals to a hitherto unconfirmed sense of a shared social practice were often made in response to our questioning. The trouble, of course, was that, because they had often given up on thinking about their washing routines, such appeals could suddenly feel as if they were on rather shaky ground (and especially so, when what others were doing with their own private washing practices was often unknown).

We saw this, I think, particularly at the first of our two festivals. This was, I think, for a few reasons. First, attendees there were a little older, and that made them less willing to immediately give up on ideas about maintaining appropriate standards in how they spoke of themselves. It was also probably because we too were yet to slot more fully into the festival spirit of easy open discussion between erstwhile strangers. Because of this our questions could come across as more like tests from outside at first – something more accusatory than we might have ideally wanted in terms of dealing with departures from 'the norm'.

For example, a former playschool worker who was otherwise very interested in thinking about festival dirt (and particularly in what might be going on, based on the available evidence, in some of the less well-kept festival tents she'd noticed over the years), was, for these reasons, less forthcoming with views on the cultural importance of everyday washing. She expressed this as 'in my head, everyone does what is normal'. In doing so, she was both positioning the actions of others as similar to hers and putting them beyond reproach. However, at the same time, she was unsure of what they were. So, hers was quite a hopeful positioning. And, in that respect, her view was similar to the idea of 'liking to think' they were 'fairly hygienic' as expressed by Kate – they didn't really know what others were actually up to at home, but it was nice to see themselves as living in line with appropriate standards (partly because doing so meant they didn't have to think any more about it).

Gaining knowledge of, and comparing your own actions to, the washing strategies at large in wider society emerged as quite a fraught undertaking. Indeed, it was easier to live with certain half-developed thoughts about being 'fairly hygienic'. Another respondent, when asked how about she felt about the idea of being personally dirty in everyday life back home, snapped back (with some degree of comic indignance) that she was 'no more dirty than anyone else!' In doing so, she wasn't really answering the question. And it was that feature that was most revealing here. The appeal of seeing herself as comparable to others was so strong that it trumped our question. It was more important to convey a sense of how she was similar than to air any of her own 'dirty secrets' by talking of how she felt about being dirty.

This simultaneous interest in what others were doing with their washing and their limited access to it was nicely encapsulated in an anecdote (in which, of course, the fact that it was sufficiently 'remarkable' to become an anecdote was already telling). Ellen described a series of experiences with housemates during her university years that illustrated how this matter was collectively managed. Crucially, these were housemates that she didn't know especially well and so, though they lived 'together' after a fashion, they didn't discuss certain things. One of these was personal washing. For washing, comparison came through more subtle means. As a short person, she worried about the fact that she moved the showerhead down each time she used it. Would that give the game away in terms of how often she washed herself? She didn't think that she showered too infrequently. But that was beside the point. The point was that she didn't want them to know! However, it eventually became apparent that they did. The relationship with one flatmate turned sour and a formerly taboo topic suddenly provided a new opportunity for criticism. He had nothing to lose at this point. So why not reveal how he literally 'had the dirt' on her (as an expression which directly alludes to the idea of knowing something about others that they'd rather you didn't)? When he eventually left (and apparently as a kind of parting shot), he spoke of his disdain for how others went about their washing there (they didn't do enough, and he had noticed). Again, the detail of their differences was beside the point (much as it might be interesting to compare). The point that her anecdote alluded to was more about how they liked to be thought of as upholding certain standards in everyday life but seldom spoke of these matters (in this case, not until a low-level friendship eventually fell apart).

And so it was, in this fourth project more than the other three, that we saw anxieties about discussing everyday practices with us. More than once respondents spoke of 'the person who might be listening to the tape' (with reference to the Dictaphone that had been placed between us to pick up our talk without that being obscured by the various festival songs and sounds). In other words, they were particularly alive to how our festival conversations were being recorded and nervous of how some unknown others might feasibly judge them in the future. Perhaps this was to do with their generational experience and how their heavy social media usage may have made this group particularly alive to these matters. It may also have related to how, as described above, in the sanctuary of the festival there was a sanction to mull over matters that they probably wouldn't discuss elsewhere. But this was also because, in the absence of complete knowledge about wider showering practices, they wanted to think of themselves as essentially the same as others (but were unsure of whether they really were). As I discuss next, that also made the festival experience particularly interesting because, within it, they were granted new levels of access to how others were washing.

A Form of Focused Relaxation

In the above account, certain ways of talking revealed how, though our festival goers had seldom given it much thought, or indeed knew a great deal about what others were doing, these young people figured (often hopefully) that they were generally doing something similar to their generational peers with their washing. But now they were at the festival. And now they could compare previously private practices. Most didn't shower there. This was admittedly partly because the shower blocks were few. This meant showering involved quite a bit of queueing (though that, in itself, could be a bonding experience). But it was still a strange situation. So, how did they deal with doing something very different with their personal washing whilst also witnessing how others handled this feature of the festival?

One way of coping was to emphasise the situated nature of their actions, both to us and to themselves – how in everyday life things were structured so that they washed a lot and, now (as a result of finding themselves in a very different environment) they no longer did. As simple as that! But this step change in personal washing practice also involved some degree of wilfulness (even though the festival atmosphere, as will be discussed, encouraged them to present this to us and others as effortless) – a determined suspension of usual washing practices. More than other holidays or leisure trips away from everyday life, the festival was explicitly about being different – and that required an interesting purposefulness to their apparent purposelessness in ignoring the shower.

Meanwhile, not washing in everyday life could 'just feel a bit wrong somehow' according to Polly. Though on occasion, she didn't shower, and readily admitted to enjoying the 'warm snuggly feeling' that could come from that (like the students who worried about sweat in the above Australian study), she only did this privately at home. Then, at the festival, she felt that 'you've got to get a new cleanliness mindset' and you 'need to get your head around it'. The language here was deliberate – for her, this wasn't an effortless descent into a pleasurable experience of collective decoupling from 'standards'. This was about pushing through a temporary disquiet when something that was usually done without thought was brought into her consciousness. Though not showering eventually became enjoyable (and was easier than they expected), she, and others, had to hold their nerve to get there.

A Rebel without a Cause

In some of our interviews, as I said earlier, others joined in with the discussion. On these occasions, our talk would naturally shift from personal reflection to collective evaluation. At times, this could feel jarring when we had been working hard to establish the right kind of confessional tone for exploring private matters with a comparative stranger in a relatively public place. But, on a more positive note, it also offered an opportunity for a different kind of analysis. Usually this happened mid-way through the interview when the friends for whom the respondent was waiting eventually arrived. At these times, we found ourselves bearing witness to how a group of friends talked about festival cleanliness. And that also proved instructive in learning about how talk patterns and personal practices connect.

'You're a rebel without a cause!' chipped in one friend as our respondent spoke of how he was happy not to shower for a few days at the festival. They had piped up with something similar earlier (when he had jokingly swiped their comment aside with an ironic deflection – saying that it was indeed 'really hard' for him to do so). They were teasing each other. But it was still true that, though they collectively were of the view that it 'really should' be easy to go without washing over the festival lifetime, it turned out to be harder than he anticipated (though, importantly, he, like most, hadn't actually done too much 'anticipating' about these aspects). The comedy rested on the mismatch between the perceived challenge of the festival for him – which was not insignificant – and how going without a shower for three days 'really should' be easy (such that making reference to movies about rebellion was deliberately ridiculous). Elsewhere in our project we talked with our respondents about what one of them dubbed their festival 'holiday from hygiene' (an engaging term that seemed to us, at the time, to capture their experience). But, for many, there wasn't always quite this kind of easy embrace of a different way of relating to bodies, environments, societal conventions and outdoor experiences. Rather such phrases belonged to a broader exercise in bolstering an upbeat way in which experimentation was encouraged by the temporary festival community that had started to emerge. In other words, they were acting into the social context as much as relaying the truth of what people felt. As will be considered next, they were serving to create, as much as reflect, their experience.

Prudence also attempted to make light of these challenges. Sitting in a field of long grass next to her tent (artfully selected because it was nicely away from the buzz of the stages and the food stalls), we are working through her routines. For an otherwise articulate person, who appeared to possess a strong sense of why she did certain things in everyday life, her usual showering routine had comparatively few justifications attached to it. Eventually, she exclaimed – 'what is this neurotic grip that has gotten hold of me?' The point to take away from this exclamation was the same as that linked to the 'rebel without a cause' comment. Escaping routines should be easy. Yet, somehow, it was hard. Still, and as will be discussed next, certain ways of talking could help them to push through these unsettling feelings.

All in the Same Boat

Many of our respondents talked about being suddenly presented with a radically new situation in terms of relating to dirt, their bodies and being outdoors at the festival. And that was true. But it also alluded to the limited amounts of prior thought that had been devoted to this matter. Elsewhere in our study, we asked attendees to complete a survey for us. In this we found that, though many of them brought a towel to the festival, very many of them found afterwards that they didn't use it. In other words, they had prepared based on a relatively unthinking presumption that it would be helpful (they always showered elsewhere in their lives, so why would this situation be any different?). But then they didn't. The idea of the 'unloved towel' was worth thinking about (and it was about more than the limited numbers of shower blocks at each festival). How had it come to pass that it was so often left alone and unloved? Theirs wasn't, as discussed, always an immediate and easy transition into towel indifference. For example, the point in the morning when many people usually washed was an especially unsettling time when routines were to be re-examined (even though those who reproduced them were often both half-asleep and hungover).

Certain festival features helped. One of these related to how they were staying in fields of tents that blurred the boundaries between their public and private lives. This meant that cues could be taken from observing actions that otherwise happened behind closed doors. Indeed, for some, this was charming, almost life affirming, as they noted, in passing, that everyone faced the same everyday problems (irrespective of the perceived pressure to 'present' something more polished when people leave home and set out into the wider social world). Either way, watching instances of others not doing very much in the morning, or perhaps even having a can of lager to start the day, could 'take the pressure off'. But that was not to say that this was a straightforward process of putting the usual practice aside.

'it's really not cool to shout out "Oh no, I just spilt some lager on myself."'

The above quote provided a good example of how certain norms of acceptable speech at the festival helped keep new ways of relating to their bodies in place. I'm back with Jenny, who is still in a reflective mood as we are looking down onto the main festival area. The festival is warming up as the interview went on, with queues beginning to snake out from the breakfast stalls below. In view of how the festival was determinedly positioned, both by her and by most people with whom we spoke, as an accepting place in which personal judgement was rare, this was an uncharacteristically strong opinion from her. I was a little taken aback by it at the time. Her stance on this matter was definitive. There was no debate to be had here. Indeed, she was more accepting of those who peed in unusual spots at the festival than she was about such public exclamations about becoming 'unclean'.

Spilling lager on yourself was evidently just how things sometimes went at the festival. It was something to be expected. But her remembered annoyance at overhearing a public expression of anxiousness about this (she was recalling a past experience in the above quote) was also about saying it aloud. The problem was that exclaiming about having some lager on yourself might upset the delicate balance of socially acceptable escape from usual standards that had begun to develop there (but was still quite fragile). And so it was annoying because it risked others snapping back into an awareness of (and potential anxiety about) how they were no longer washing so much at the festival site (especially annoying when everyone was working hard to forget about this!). As David put it, it was 'nice not to worry about showering'. Such exclamations could make these worries suddenly flood back.

Yet, whilst certain public forms of speech could unsettle newly acquired ways of relating to their bodies (and were therefore unwelcome in the collective), others helped these new relationships to become more fixed (and encouraged others to embrace them too). As mentioned, at both our festivals, the provision of showers was patchy (there were very few there in comparison to the total number of attendees, and festival shower blocks were often tucked away at the edge of the site – out of sight and out of mind). But what was particularly interesting with regard to this was how this provision was talked about amongst festival goers. 'Have you seen the length of the shower queues!?' was a frequent refrain, for example. On the one hand, this could run the risk of the above 'lager exclamation' in the sense that it reminded others about wider washing imperatives. But, on the other hand, and as was more commonly observed, it could have the opposite effect. In other words, it provided a legitimating narrative for not using the showers. This was doing something more than simply conveying useful information about on-site facilities between erstwhile strangers. It was allowing them to come to a collective decision about letting standards slip.

'We are dirty and we don't care!' Such was the proud chant of a pair of women in their thirties that we encountered sitting on a bale of hay one morning. And they certainly were. They had been up all night and were quite happy to recall 'rolling around' together a few hours earlier. But they were also serving to solidify a particular idea about how you might properly 'get into' the festival experience. Their public exclamation was helping to shape how festivals were to be understood by the group (they were doing more than sharing their private thoughts and opinions with us). Meanwhile, those who found these shifts more of a struggle, spoke in a comparatively hushed, confessional tone. There was a general sense of acceptance, even gossipy enjoyment, about how private practices were now subject to public debate in ways that facilitated various forms of bonding. And that, in fact, could make it difficult for those who stoically stuck to their usual washing practices. One advertising manager confided in us (in more of a muted tone than that he had adopted up until this point) about the infrequent showering that he had observed there:

'and people say it's part of the experience, but deep down nobody likes that.'

His use of 'deep down' suggested the degree to which prevailing washing practices had become so internalised by society for him (they had taken root 'deep down' and 'nobody' felt otherwise) that the festival could never really disrupt them. It also seemed to indicate how certain things were less available for discussion there (some things were readily offered up in the interview; others required careful excavation and more conspiratorial exchanges about less publicly sanctioned opinions). But, for the moment, the point was about how, as within our interviews, certain forms of speaking predominated at these festivals. An upbeat claim that they were all 'in the same boat' was commonplace at both festivals. In saying that, attendees weren't just describing their situation – they were making it too.

This presented a problem for those who were staying in the 'elite sections' found at both festivals – the 'glamping' areas in which were found various additional features including (of particular interest to this chapter) privileged access to extra showers that were also cleaned. This led them to wash differently to many of the others. But they weren't especially proud about that (because taking advantage of them was not in line with the 'all in the same boat' ethic). When we spoke with some of these 'glampers' they often preferred to talk about the shared challenges of comparatively simple living at the festival. Doing so gave them access to the 'same boat' narrative that was valued by almost all the respondents as an important aspect of what makes the festival special (whilst also quietly enjoying their extra facilities).

All the Tree Stuff and the Flaky Bits

Compared to the three preceding chapters, in which the conversations (or, more rightly in terms of my focus in this book, the patterns of talk that could be observed and analysed) were often nudged back to consider the benefits of being in and around plants and trees, this happened less often at the festival. That is partly because the ambition this time was to stay focused on issues to do with the physical experience, and on new ways of washing. These issues seemed much more pressing than a background sense that, for example, the trees behind the stage were 'nice things to have'. But that is not to say that being in rural greenspaces outside did not shape how these respondents spoke of the experience.

For some, this was definitely an afterthought. Indeed, at times, it seemed to come as a shock when the suggestion of being in a more 'natural' environment was introduced by us (an unwelcome conversational insertion when animated gossip about personal washing was previously being enjoyed). Some even refused the suggestion ('oh, it's just grass really'). The various distractions of the festival definitely trumped this topic. Others offered up some relatively banal statements with little apparent passion – 'yes, it's a lovely thing to do' said one; another was hesitant about sounding clichéd in saying 'it's good to get back to nature, so to speak'. Compared to the issues of how people should, or shouldn't, wash there (which piqued their interest much more), these issues were often quite quickly glossed over. Linking back to Chapter 5, this was partly because such matters seemed difficult to speak of in ways that went beyond the rehearsal of statements about personal benefits that many felt sheepish about making (recall the reticence about experiencing the garden in a 'mad Prince Charles way'). It was also because these features were far from their minds.

Exploring these issues required a little more thinking. And so we asked how 'an urban festival' might be different. These wouldn't, so it seemed, be the same thing at all! Certainly this environment offered a feeling of escape and that did, in turn, filter through into how easily it was for them to relate to their washing in new ways. Crucially, it allowed them, in part, to embrace what was taken to be a more 'child-like' state (just as some of the office workers spoke of how certain peers responded to dramatic weather) – they were away, and they were 'in nature', and that meant more unprescribed forms of enjoyment as they were temporarily allowed to become 'cave men in make-up'. And so, just as it was in some of the earlier chapters, the natural world became something that was properly for children. It was not really part of the routine life of the modern professional person who had things to be getting on with and who was demonstrably part of an appropriately civilised urban society. In fact, that was part of the pleasure of their spell in what David called an 'adult playpen.'

It also made for a different physical experience. Again, that was often easier to discuss when compared to feeling forced to make vague claims about how it was 'nice to be near trees'. The air, for example, was different (and that could help to allay sweat anxieties because it took all the smells away). Being outside 'made you feel more natural' too, for some. The sun could also be taken to dry up any sweat that you may have produced. So there were pragmatic benefits to this. It was also hard to speak of how periods without washing might be received differently were they in the city. This was difficult partly because back home they embodied a routine (and that meant active reflection about certain issues was unrequired). However, some faltering statements gradually came through. Not washing back home would leave them feeling 'ucky', 'gross', 'unkempt' (along with various other evocative terms that we had not often heard before). And these embodied states were both unpleasant and hitherto unspoken. Meanwhile, a 'different kind of dirt' was found at the summer music festival – a special kind of dirt that might even be personally improving – like a 'mudpack in the forest' (Ellen). So being around 'all the tree stuff and the flaky bits' (Tim) was sometimes annoying. But it was definitely also a nicer kind of 'stuff' than that which was found outdoors in the city. Discussing how 'nice' it was to be outdoors in greenspace wasn't an especially attractive proposition. Talking of an environment in which sweat could dry, smells were blown away and the dirt was somehow different was much easier.

Should There Be More Showers at the Festival?

I'm sitting on yet another patch of grass. This one is more public that I'd probably like. It's immediately alongside the main path between the food area and a wooded expanse in which there are various activities and exhibits for people to engage with when they feel they need some rest and respite from the intensity of the main festival. The idea of a late afternoon break seems popular today and there are plenty of people walking around this patch. I'm vaguely worried that someone might step on my Dictaphone! I'm talking with Geoff, a junior solicitor who had been enjoying exactly this sort of break moments before. He was going to meet some friends later. But he also liked the sound of our study and so he said he was happy to do an interview with me – so long as it 'didn't last too long' (nicely unspecific, I thought). So, we sat down and started talking there and then on the verge.

Like most of those with whom we spoke, Geoff did quite a bit of showering. In fact, he showered more than most. Geoff took up to three showers every day back home – partly because he 'needed to' after sport, partly because showers could somehow (though he hadn't thought too much about how exactly) 'reduce stress' and partly because, if he didn't shower at certain times, he would feel like he had been 'taken off track' such that he wouldn't 'perform' so well at work and in life. I linked this to the festival and that led him straightforwardly to suggest that, as a result of how he relied on showers in everyday life back home, it would be very welcome if festival organisers also provided more.

The conversation then evolved into a more rounded consideration of his festival experience – how it was 'actually nice' not to worry about appropriate self-presentation as he did in everyday life back in London, and how putting such worries aside helped people to relate to one another on a more enjoyably 'simple' (as he put it) level. As this process went along, he found himself coming to take a very different position. Now his thinking was that he didn't really want more showers there. They would tempt him back to them. And that would be a bad thing if part of the pleasure was about doing something different, about relating to others in less judgemental ways such that no one needed to do all that washing after all. He was a little taken aback by that. We considered the contradiction in silence for a second. Eventually, he added in a more contemplative tone – 'I guess I love what I am used to.'

On one level, I was arguably steering the conversation – bundling him onto two quite different trains of thought, each with a different final destination in terms of his apparent opinion. Nonetheless, when we discussed in a relatively abstract way whether there should be more showers provided to festival goers, the answer was apparently straightforward. In view of how tickets were expensive, and easy shower access was unthinkingly taken to be a positive thing irrespective of context, the answer given by almost all of our respondents was an immediate and unequivocal 'yes'. Why would they argue against this when showering, in everyday life at least, was understood as necessary, pleasurable and (though they often couldn't quite articulate why) somehow definitely 'important'? Yet, as our discussions gradually moved towards the enjoyment of escapism, collective communion and opportunities for suspending social norms, answers drifted towards the opposite position. At this point, respondents began to realise how they had started to assume a different stance. Showers were evidently good. But going without them could be good too.

Indeed, going without them might, on further reflection, 'even be part of the attraction of it' (David), as another seasoned festival goer ended up concluding. This transition was literally embodied by Prudence, as one of the most reflective of the respondents in this study. Towards the start of her festival experience she did what she described as 'an instinctive shower reccy' (reconnaissance exercise) because that is what she assumed she should do (because she assumed that she'd be using them there). Then slowly she came to realise that she didn't need or want to shower and laughed at herself for her 'unthinking reccy' (which was rather like the unloved towel earlier on – a matter of habit more than desire).

The Return of 'The Office'

So far, we have seen how usual ways of relating to water and personal washing were suspended at the summer music festival. But it would be wrong to leave it without acknowledging how this was, without doubt, no more than a temporary suspension. Our respondents wouldn't do anything different with their personal washing because of their festival experiences once they had returned home. All-too-familiar social expectations would soon fall back upon them and usual showering routines would accordingly be re-established. They had no choice about that. One respondent spoke towards the end of the experience about how she was already anticipating feeling 'grubby' on the train back to London. She didn't feel that way here. But being surrounded by others who had been washing much more would make her wince with self-awareness. What was also noticeable was how 'the office' featured as a commonplace shorthand for conventional life back in the city. They were subject to certain expectations in 'the office' and we all knew what they were. And so, in one way, we've now gone full circle as some of the sentiments expressed in the first case study in Chapter 3 reared their heads once again. Whether individually they liked it or not, 'the office' idea brought with it a recognised set of pressures and social precedents that no-one thought they could challenge because they were so entrenched. Once again, and just as was found in Chapter 3, 'the office' could mean doing what was expected and embodying the 'norm' – it was a place set up for conformity and for forgetting about doing otherwise.

Conclusion

Embracing the Outdoors

What should we take from this fourth case study in terms of the first wager? What has been learnt about how the outdoors can prove unsettling to certain otherwise entrenched ways of personal washing amongst young Britons? First, we learnt that it was less unsettling than might have been expected. Despite the evident level of attachment to, internalisation of or perhaps even (depending on how you look at it) surrender to a practice of frequent washing that was hopefully positioned to us as widespread, escaping it wasn't so problematic. For those who would encourage us to live otherwise with the outdoors, with our bodies and with water more generally, we could therefore see quite a positive story here. We could tell a story of how those who had been fairly fully subsumed into a perceived culture of frequent personal washing did not find the experience of desisting to be especially disgusting. Certainly, ideas about feeling 'gross' were less common than we anticipated. In fact, we concluded that we might see festival goers as 'creatures of context' in the sense that, so long as others were evidently doing something different with their personal washing, they would do it too. Furthermore, doing less was not really so onerous, and sometimes even felt like an enjoyable return to a more authentic human state.

In terms of what we might do with this finding, it is not as simple as giving people what they 'want' when our 'wants' are contextually produced rather than being a set of dearly held preferences that are fixed and unwavering. This made us conclude that the ways in which those who decide how to take 'outdoor hospitality' forward from this point have an important hand in 'making' those who they welcome, rather than being mere slaves to apparent demand. And so one of the suggestions that came out of the study related to how a practice of routine and regular showering can be more easily escaped than might have been anticipated. Of course, that is partly the point of a trip away since they are designed to provide something different. But these festival goers hadn't gone there with a sense of excitement about living in more basic ways. They were not particularly seeking a situation that could have been seen as hardship. It was sometimes unsettling in the mornings, when they saw a diversity of responses that made them unsure about how they should act themselves. But change still came. Conceptually, the implication is that we should not overplay internalised routine and, practically, we can take heart from the idea that if we put people in a different environment, and others seem to be handling it, that could turn out to be enjoyable. Returning to the issues that began this chapter, this presents interesting challenges for festival organisers and others who have been eyeing up the 'glamping market'. They should not assume that customers always have predetermined requirements. They may very well develop them if increasingly cocooned from the challenges of experiencing anything that dips below the levels of service to which they are elsewhere accustomed. But something may be lost in the process since pleasure can also come from escaping the expectations produced by infrastructures that ironically only seek to please back in everyday city life.

Having said that, the apparent benefits of being in greenspace were seldom explicitly part of this for people (though they thought assuming this alternative state would definitely be harder in urban contexts). There is a mixed picture here in terms of what this means for those (like myself) who are keen to safeguard and encourage the restorative benefits of being in and around plants and trees more broadly. On the one hand, there was another positive story here about how, if we see comparatively simple living outdoors with 'nature' as good (as those who promote wilderness therapies and other strategies do), those who face challenges to their usual ways of living in embarking on such exercises can definitely adjust. However, on the other hand, seeking out and potentially deriving these benefits may not be assumed to be a natural by-product of putting them in these situations. These respondents were definitely in and around trees. And they were also camping in fields. But the extent to which that was considered (or rather found its way into their way of thinking about what was happening to them there) was another matter. It is quite possible of course that this doesn't need to register with them – that vegetation acts as a background source of wellbeing that charges people up (and provides benefits without them particularly considering this process). If so, the extent to which it registers is neither here nor there.

Either way, and turning to the final way in which I would characterise how specific relations with the outdoors developed for those in each of the four groups that I studied, in this case I saw the outdoors being eventually 'embraced'. By starting my analysis with the practice of frequent showering, we saw how that practice was suspended for a period and replaced by a different way of relating to bodies outdoors. Though this was sometimes quite gingerly done (as usual practice was put aside), and though it was often carefully developed by the collective, new ways of being and living in outdoor spaces were definitely seen to emerge in my final study. This is, however, not to say that these young people felt they wanted to live their lives all the time as they did at the festival. But it is to say that escaping certain everyday practices and embracing another way of doing things is far from impossible and, to end on a positive note, can (so long as we know that others are doing it too) appeal. In that regard, the young (generally urban) people involved in this case study were not so used to being close to their showers that they couldn't embrace a different way of living amongst green outdoor environments. But care was needed to effect such an embrace too.

Speaking of Showering

I return now for one last time to my three suggestions about how people are likely to speak of otherwise routine social practices. This final context was quite specific since in this study we were able to observe and enter relevant conversations that were already happening (as well as organising our own interviews). So, what has been learnt here?

- The first was about how practices encourage particular ways of talking. In this case, there was little evidence of any patterns of everyday speech that actively kept people in line with what was nonetheless a relatively uniform practice of frequent showering. This was a private matter or rather an unspoken practice. Perhaps the practice no longer had any need of such speech patterns in that it was already so well established that people were no longer required to rehearse them. There may be occasional verbal nudges in shower product commercials or public health campaigns perhaps, but they were not readily available to people when examining their everyday actions. This final case therefore lends weight to the suggestion that sometimes practices can become stronger through the situated disappearance of talk and reflection, though, of course, I could only become sure about this by seeing whether any showering 'chat' was available to those with whom we spoke.

- That first finding could lead to anxieties coming to the surface when being asked about these matters (perhaps in a particular way, though not especially, I think, as a result of us being erstwhile strangers). When people, for example, made reference to our recordings this suggested a greater level of interest in (and potential anxiety about) what would happen with their responses than was seen in the other three studies (in which people more often dived directly into the discussion). Jokes featured here too, though this was in a slightly different way to the idea that they neutralise anxieties about idiosyncrasy. Here their use was less about diffusing worries, less about acting out of line with what others were doing and more about dealing with the realisation that you didn't know quite why you did particular things. Accordingly, in the context of this absence of prior thought, certain question sequences could prove especially powerful. As we saw, when showers were presented as a facility that could be offered, they were wanted. But when discussion circled towards the pleasures of embracing the outdoors, they were no longer so necessary. So, in the context of a comparative lack of prior thought, we can learn a lot from embarking on such journeys of self-discovery within our interviews.

- Finally, in terms of how appeals to the collective came about, these were again in evidence. But the circumstances were a little different here. When 'I'm the same as everyone else' is advanced as forcefully as it was at the festival (I don't think the office workers, or the garden shoppers, or the runners in either environment would be so emphatic), a background sense of common practice was being summoned to push through a passing anxiety. Being recognised as toeing the line with a wider convention was perhaps all the more important if you hadn't thought much about it and worries about potential social judgement suddenly surfaced. But what was perhaps most interesting (and something to which I had not been granted the same level of access in the other three studies) was how new collectives were made through certain forms of speech (not actively 'made' in the sense of being deliberate, but made nonetheless). This was seen in claims about how 'no one washes at the festival' or when certain ways of speaking were deemed 'not cool' when they imported outside anxieties that were unwelcome at the festival. And that was little to do with what happened in everyday life and much more about how fledgling forms of speech helped the festival collective to embrace an alternative.

References

Andersen, J. (2013). The music festival – communicative recognition with a political aspect. Paper for 7th ECPR General Conference Sciences Po, Bordeaux, 4–7 September.

Anderton, C. (2008). Commercialising the carnivalesque: the V festival and image/risk management. *Event Management* 12: 39–51.

Brochado, A. and Pereira, C. (2017). Comfortable experiences in nature accommodation: perceived service quality in glamping. *Journal of Outdoor Recreation and Tourism* 17: 77–83.

Brooker, E. and Joppe, M. (2013). Trends in camping and outdoor hospitality – an international review. *Journal of Outdoor Recreation and Tourism* 3–4: 1–6.

Browne, A., Jack, T., and Hitchings, R. (2019). 'Already existing' sustainability experiments: lessons on water demand, cleanliness practices and climate adaptation from the UK camping music festival. *Geoforum* 103: 16–25.

Browne, K. (2009). Naked and dirty: rethinking (not) attending festivals. *Journal of Tourism and Cultural Change* 7: 115–132.

Bultena, G. and Klessig, L. (1969). Satisfaction in camping: a conceptualization and guide to social research. *Journal of Leisure Research* 1: 348–354.

Collado, S., Staats, H., and Corraliza, J. (2013). Experiencing nature in children's summer camps: affective, cognitive and behavioural consequences. *Journal of Environmental Psychology* 33: 37–44.

Cordell, H. (2008). The latest on trends in nature-based outdoor recreation. *Forest History Today* Spring: 4–10.

Flinn, J. and Frew, M. (2014). Glastonbury: managing the mystification of festivity. *Leisure Studies* 33: 418–433.

Garst, B., Williams, D., and Roggenbuck, J. (2009). Exploring early twenty-first century developed forest camping experiences and meanings. *Leisure Sciences* 32: 90–107.

Getz, D. (2010). The nature and scope of festival studies. *International Journal of Event Management Research* 5: 1–47.

Gram-Hanssen, K. (2007). Teenage consumption of cleanliness: how to make it sustainable? *Sustainability: Science, Practice and Policy* 3: 1–9.

Gram-Hanssen, K., Cristensen, T., Madsen, L. et al. (2019). Sequence of practices in personal and societal rhythms – showering as a case. *Time and Society* 29: 256–281.

Hand, M., Shove, E., and Southerton, D. (2005). Explaining showering: a discussion of the material, conventional, and temporal dimensions of practice. *Sociological Research Online* 10: 2.

Jaimangal-Jones, D., Pritchard, A., and Morgan, N. (2010). Going the distance: locating journey, liminality and rites of passage in dance music experiences. *Leisure Studies* 29: 253–268.

Lea, J. (2006). Experiencing festival bodies: connecting massage and wellness. *Tourism Recreation Research* 31: 57–66.

Louv, R. (2005). *Last Child in the Woods: Saving Our Children from Nature-deficit Disorder*. Chapel Hill, NC: Algonquin Books.

McKay, I. (1994). *Quest of the Folk: Antimodernism and Cultural Selection in the Twentieth-Century*. Montreal: McGill-Queens University Press.

MINTEL. (2018a). *Camping and Caravans*. London: MINTEL.

MINTEL. (2018b). *Music Concerts and Festivals – UK*. London: MINTEL.

Morris, N. (2003). Health, well-being and open space literature review. OPENspace: the research centre for inclusive access to outdoor environments. Edinburgh: Edinburgh College of Art and Heriot-Watt University.

O'Neill, M., Riscinto-Kozub, A., and Van Hyfte, M. (2010). Defining visitor satisfaction in the context of camping-oriented nature-based tourism – the driving force of quality. *Journal of Vacation Marketing* 16: 141–156.

O'Rourke, S., Irwin, D., and Straker, J. (2011). Dancing to sustainable tunes: an exploration of music festivals and sustainable practices in Aotearoa. *Annals of Leisure Research* 14: 341–354.

Pretty, J. (2004). How nature contributes to mental and physical health. *Spirituality and Health International* 5: 68–78.

Pullinger, M., Browne, A., Anderson, B. et al. (2013). Patterns of water: the water related practices of households in southern England, and their influence on water consumption and demand management. Final project report for the Economic and Social Research Council.

Purdue, D., Durrschmidt, P., Jowers, P. et al. (1997). DIY culture and extended milieux: LETS, veggie boxes and festivals. *Sociological Review* 45 (4): 645–667.

Russell, H. (2001). What is wilderness therapy? *The Journal of Experiential Education* 24: 70–79.

Strengers, Y. (2009). Bridging the divide between resource management and everyday life: smart metering, comfort and cleanliness (PhD). RMIT University, Melbourne.

UK Festivals and Awards Conference. (2018). 2018 market report. http://www.festivalawards.com/insights (accessed 20 April 2021).

Waitt, G. (2014). Bodies that sweat: the affective responses of young women in Wollongong, New South Wales. *Gender, Place and Culture* 21: 666–682.

Wells, N. and Lekies, K. (2006). Nature and the life course: pathways from childhood nature experiences to adult environmentalism. *Children, Youth and Environments* 16: 1–24.

Chapter Seven
Conclusion

This book began with two wagers.

The first was about the ways in which greenspace experiences, along with the various human benefits that appear to be associated with them, are accommodated within different domains of contemporary life. The wager was that, whilst there is good reason to undertake studies that identify, explore and emphasise these benefits, without a sense of how relevant patterns of everyday action become entrenched in societies, we are missing an important part of the puzzle. Indeed, I argued that, if we want to identify the most effective strategies for promoting the many apparent positives of spending time in the company of trees and plants, we might benefit from turning to how particular social practices could, in effect (and often without them especially noticing at the time), be driving some increasingly sturdy wedges between people and greenspace. Based on that suggestion, I argued that it was worth looking at how such processes are playing out within some of the everyday lives involved. The idea in essence was to look at how particular outdoor experiences that would likely involve some time in the company of living vegetation, may, in effect, serve to destabilise certain everyday practices (and, by association, the people who carry them out) that are currently common (and sometimes increasingly widespread). In other words, the idea was that, though 'the outdoors' may very well be 'great' in principle, we might learn something from exploring how it can also be 'unsettling' in practice.

The Unsettling Outdoors: Environmental Estrangement in Everyday Life, First Edition. Russell Hitchings.
© 2021 Royal Geographical Society (with the Institute of British Geographers). Published 2021 by John Wiley & Sons Ltd.

The second was that, in order to shed a useful light upon these processes of 'environmental estrangement', we might benefit from paying attention to how relevant groups of people speak of some of the practices with which they are familiar. This may seem like a fairly uncontroversial way of going about such an enquiry. But, as I discussed earlier, doing so can also run against the supposition that people are either unable or unwilling to speak of the practices they are routinely involved in reproducing. For some, they were thought to be unable because their practices have become so ingrained as an unthinking response to certain situations that they no longer can. For others, they were unwilling because that would require them to recognise how much of their lives goes on without them making conscious decisions when they might not like to think of themselves in that way. Whilst there is some truth to both claims, rather than dismissing such an undertaking at the outset, I took this as a spur for a closer examination of what could be learnt from how people spoke of certain otherwise often relatively automatic actions. The second aim of this book was therefore to use my studies to identify some of the exchanges and responses that others might look out for and explore when speaking to people about their practices. In this regard, my idea was that, even though my own projects were focused on outdoor environments, my experiences in talking about the everyday lives of others stood to provide some more general insights about studying social practices through spoken exchange.

With these two wagers made, I turned to four studies with the potential to illuminate both matters. And we have now travelled to various places in which those found there were asked to speak of how particular ways of going on could make certain outdoor experiences an unsettling prospect. The journey started with some professionals found inside the corporate offices of central London. In some senses they were those who were most effectively protected, depending on how you look at it, from ever having to trouble themselves with thoughts about the outdoors in their working lives. Then we ventured inside another environment in which many of these office workers could also often be found, namely the gym. Here we saw instances of recreational running, as a practice that many of us may still like to associate with outdoor environments, being quietly relocated onto indoor treadmills that were practical and, in some important senses, addictive, even if they were, at the same time, quite far removed from the imagined ideal. After that, we turned to the domestic context. The home is a place in which we might imagine there is more time and inclination to engage more fully with the outdoors for those lucky enough to own a garden. Here I explored how this was not always so easy because, for some of them, the idea of waiting, of growing or even of learning about the plants that were found there was an exciting suggestion in principle, but quite difficult to implement in practice. Finally, we took a short trip outside the city to see how young people, as a group that spends more time washing themselves than most, responded to basic showering facilities and the prospect of fairly simple outdoor living at the summer music festival.

They were somewhat startled by the alternative mode of managing their bodies that they found there. But anxieties were eventually overcome, if only for a reassuringly finite period.

We have met a lot of people along the way. Sometimes with colleagues, I've squatted down in fields in order to speak with a stranger who was, at the time, coming to terms with comparatively primitive living conditions; I've become hot and bothered as I tried to pose effective and acceptable questions whilst keeping pace with the person running alongside me; I've been anxious about seeming sufficiently serious as a researcher whilst asking those who are otherwise busy to reflect on how often they look out of their workplace windows; and I've sat on patios with those who would like to have a lot to say about the gardens on which they have spent their money, but still sometimes found themselves comparatively lost for words. Though there were definitely challenges, I think that asking all these people to speak of these experiences produced useful insights. Indeed, the time has now come for me to return to my two wagers. Enough material has now been examined to say something to both those interested in the social trends that could be serving to separate people from the outdoors and those curious about strategies for taking an interest in the everyday lives of others.

Wager One – on Processes of Environmental Estrangement

A Language of Lived Extinction

I begin this first discussion by briefly returning to how I am drawing on social theory in this book. As discussed at the start, I was particularly attracted to the idea of how identified 'social practices' take hold of people. Part of my aim was to acknowledge that, despite how culturally ingrained and individually attractive it can be to think of people as generally in control of their actions, we are often quite malleable creatures who think about what we are doing sometimes, but not always, and who are vulnerable to how practices can capture us and control our actions thereafter. Whilst this may seem a misanthropic way of seeing social life, my idea was that no-one wants always to be thinking about the detail of what they are doing, whether they are doing it right or whether they could do it better.

But, at the same time, I didn't want this to become an exercise in collecting examples that proved my point. To do that would be to drastically truncate the potential for learning from all those who took part in my studies. Instead, I wanted to travel light with my theory by exploring the extent to which it helped me to understand my four case study contexts. Without doubt, certain concepts were helping to train my attention onto processes that I might not have otherwise noticed. But I was also keenly aware that interview research of the type that I undertook is partly about giving the social context at hand the time that it needs to reveal its secrets. That said, I do think that all four cases point to the importance of exploring how everyday practices can present challenges to those hoping to promote greenspace benefits. After all, my basic wager was that, if we don't attend to these processes, societies could sometimes be sleepwalking towards future lifestyles in which many have gradually, and in some important ways unthinkingly, been drawn away from the benefits that these spaces can bring.

Furthermore, beyond the loss to these individuals (in terms of personal respite), this could mean, at least according to some of those with whom this book began, that many may come to care less about the health of the wider environment by virtue of their increasing disengagement from it. The suggestion here was of a self-perpetuating loop of diminishing experience leading to diminishing interest and diminishing experience once again. In this regard, we can imagine future societies becoming blithely indifferent to environmental problems outside because they have become otherwise preoccupied indoors. Adding to that another implication could be the entrenchment of some quite resource intensive lifestyles since it generally takes energy of various forms to prevent everyday lives from being too troubled by the outdoors. We can think of all the power that is required to make unchanging environmental conditions in offices and gyms, of all the materials that must be brought in to quickly build, rather than attentively grow, a domestic garden, or the significant water consumption associated with frequent showering amongst young people.[1]

Either way, and to reiterate the focus established at the start, we can make the greenspaces available to people in cities as attractive as we like. And we can continue to extol the benefits that come from spending time in these spaces. But, if many people are being effectively captured by certain ways of going on in their everyday lives that render them comparatively oblivious to, or perhaps more rightly uneasy about embracing, the benefits of these spaces, then the mounting evidence to suggest that they are good for us will be of little effect. I started with how there feasibly remains a residual evolutionary urge to seek out the vegetated spaces that may have been crucial to our development as a species. I end by noting how the power of practices can often trump the pull of these spaces even though, when asked to think and talk about it, those involved are often well aware of the benefits that spending some time with plants and trees can bring. Returning to the map that was presented in Chapter 2 (and is reproduced again below, see Figure 7.1), London boasts a lot of greenspace that should be celebrated and safeguarded as best we can. But there is more to it than provision and a focus on relevant social trends can help us to think in new ways about whether city dwellers will continue to derive greenspace benefits. This book has peered into some of the nooks and crannies of urban life to notice processes that maps cannot easily capture. But, in terms of whether and how spots of greenery, such as those depicted below, will continue to pepper social experience, they are no less important for that.

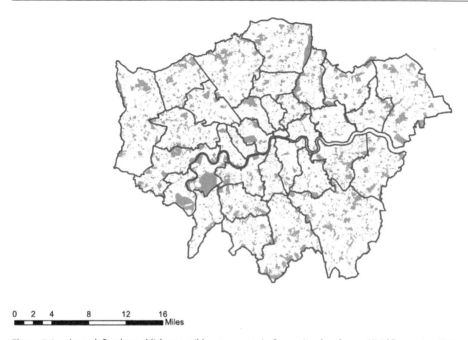

0   2   4   8   12   16
▬▬▬▬▬▬▬▬▬▬▬▬▬▬▬▬▬ Miles

Figure 7.1   Areas defined as publicly accessible green space in Greater London. Source: Vivid Economics, 2017.

In this respect, the 'extinction of experience' idea that was introduced at the start is fitting. Whilst 'extinction' gestures towards some of the heart-breaking trends in how we humans have come to handle other species, the term is also appropriate now because it hints at how relevant processes take effect slowly and in ways that may be barely perceptible at the time. We saw this in different ways in each of my four case study contexts. Within each of them, we learnt about how people could end up finding themselves distanced from certain experiences associated with being outside in spaces that are at least partly green. In the office we saw how particular bodily feelings and modes of thought could serve to keep the suggestion of having a break outside on the grass away from their working practice. In the gym we saw how the practised appeal of running was partly about no longer considering what you are doing in ways that, once you have found yourself on the treadmill, can serve to keep you there. In the garden we saw how shopping, as an activity in which very many are now quite skilled, was bleeding into the manner in which people handled these spaces in ways that can leave them with less living plants than they might otherwise have wanted. Finally, at the festival, we saw how the power of the showering practices of young Britons partly stemmed from how they had little need to reflect upon them in a way that made a brief spell of simple outdoor living in greenspaces a strange, though eventually not unenjoyable, experience. We have examined practices of office working, indoor running, garden shopping and frequent showering. All are commonplace. Some are increasingly widespread. Either way, by looking at how they took control of people, these studies provided a new window onto how the 'extinction of experience' may be practically achieved and personally embodied.

But what is more important than recognising that relevant practices can take control of people (when that finding was, to an extent, to be expected in view of how the aim was always to look out for that) is to think about how that happens in specific cases and to reflect on what can be done about that. In this respect, each chapter has attempted to characterise the social response to each suggested 'unsettling'. Did those involved retreat back towards the comforting familiarity provided by the practice at hand? Or did they, in certain circumstances, eventually escape? With that in mind, having now done these four studies, my first set of conclusions is about what to look for in terms of these lived extinctions (the processes that pull people into and out of particular relations with the outdoors). Research on the extinction of experience thesis is still developing. One of the challenges that it faces is about how to examine change over time when our ability to do so is necessarily hampered by what is known about how life went on in the past. But, if we think of these changes over shorter timescales, and in terms of how they creep into everyday lives, this book has highlighted some processes to which we could pay more attention. These processes have not often provided the focus for those hoping to foster greenspace experiences. But we probably ignore them at our peril.

Forgetting   This was most evident in the city office where we saw how easy it could be to go on without any thought of the environments outside the buildings within which those with whom I spoke were often found. This is perhaps no surprise given that the whole point of the modern office is to provide unchanging conditions that help those inside to get on with their work. But it also revealed the scale of the challenge in reminding office workers of outdoor benefits. It was also seen in the gym in which the process was, if anything, more determined. Meanwhile the garden owners could be unsettled by how the living creatures contained within their gardens required some degree of care. It was least apparent at the festival which was all about doing something different to the norm. So, it worked out differently in each of the four contexts. But a point still remains about the importance of grasping the nettle of acknowledging how easy it can be to forget about outdoor greenspace and recognising the challenge of helping people to remember when they are otherwise preoccupied.

Avoiding   This was most evident inside the gym since the whole point of treadmill running was partly about avoiding the environmental challenges of the outdoors by making the most predictable running experience for those with little time to waste. But it was also, though in different ways, evident in all four studies, as those involved effectively sidestepped, both conversationally and practically, certain potentially disruptive outdoor experiences. In some respects, avoiding the outdoors makes a lot of sense if we consider how the distracting benefits of greenspace may easily disrupt some otherwise enjoyably purposeful lives. After all, the power of these benefits, as considered by some of the greenspace studies with which the book started, relates to how they encourage us to relax, to put aside certain pressures for a second, to momentarily transcend what we are doing and potentially see ourselves as part of a process that dwarfs the problems that we face at the time. However, one important question to ask as a result of thinking in this way is whether certain groups feel they have the time to embark upon such an experience. In increasingly pressured (or, at least, what people feel to be pressured) societies, it is worth thinking more about how there will be times and places when it may be deemed best to avoid outdoor greenspaces with all their beneficial (but also unsettling) relaxation triggers.

Succumbing This process was seen most clearly in the gardens, partly because (though doing so wasn't always easy) a kind of succumbing was what some of those involved hoped to achieve within them. Many of those I spent time with during this study wanted (and liked to imagine themselves as having) a more involved garden experience. In the other three contexts, however, the idea of succumbing to an alternative relationship with the outdoor environment didn't square so easily with the idea of doing various practices effectively, efficiently and often unthinkingly. This was why the office workers preferred a half-day holiday instead of a regular lunchtime rest. This was why some runners only ran outdoors at weekends. And this was why some of the festival goers saw their ways of washing there as entirely apart from 'normal life'. But, even within the gardening case study, this wasn't always easy. And the reason it was difficult was because getting gardens efficiently organised according to aesthetic tastes under the assumption of materials being relatively predictable (or, sometimes better still, unchanging) slotted into some more familiar templates for social action. So, just as we need to think about the times when greenspace experience, with all its restorative potential, is unwelcome, so we might benefit from considering how new ways of relating to greenspaces can emerge slowly over time. Succumbing to the outdoors is challenging when it can feel risky and unfamiliar to allow that to happen and doing so can pull you out of purposeful states that are often quite valued elsewhere in your life.

Embracing This process was most evident at the festival. It happened gradually in some gardens as the designers involved found clever ways of coaxing their clients along this kind of path. It rarely happened in the running case because a big part of the idea of doing that was to be purposeful. And there was little time to embark upon such a process during the working day for the office worker. The idea alludes to the pleasures of a different embodied relation with the outdoors and a closer connection to the natural world. In the final case, we characterised this as about cultivating an enjoyable kind of indolence that involved worrying less about upholding perceived social standards. This was about young people relaxing into new embodied states even though, at the start, their existing practices could be holding them back. This process was like succumbing, but it also represented more of a short-term escape (more than, for example, a slow journey towards a different way of living in the garden) as alternative states were temporarily assumed safe in the knowledge that this was for a finite period only. This leads to questions about how long we'd want to encourage such embraces to last and the points when certain groups would likely wriggle free from them.

These are some of the terms that could help us attend to the (often unremarked upon) processes that pull people into and out of different ways of relating to outdoor environments. And they may initially seem like difficult terms to translate into further studies (Have you become estranged? Do you like to succumb to outdoor spaces?). But these are unfamiliar more than impossible to examine topics. More conceptually, the implication is that further work in this field might look out for, and examine, the times when our practices take control of us (as in the first case where those involved let thoughts of doing otherwise slip away), the times when their attributes can align with some of our own intentions (as in the second case where letting the practice take charge could also keep you healthy), the times when they are eventually escaped (as in the third case where a new way of relating to plants sometimes emerged over time) and what we can learn from instances of practice suspension (as in the final case where they were put aside for a short period). The ways in which people respond to the attributes of the practices that they carry can vary and exploring that variation can help us to understand how exactly they live through us.

Disentangled Times

On the matter of terminology, I also want to suggest that some of my geography colleagues might also make use of some of the above ideas. Though they have often been less overtly focused on promoting greenspace experience in the way that this book has been, many of them, as discussed at the start, are united by a similar passion for connecting 'people' and 'nature' in ways that could benefit all the parties involved. However, as I also discussed, for many of them, an important part of their work was about reimagining the process. In this regard, their work was partly about how we, as researchers, imaginatively embed people within the material and embodied circumstances of their lives – placing them alongside, rather than above, the various physical phenomena that we hitherto lumped together and placed under that unhelpful 'nature' banner. The problem, as they saw it, was about how the temptation to position humanity as somehow removed from these matters (as part of a strange separate collective called 'society') had been so strong. In view of that, it made sense to speak instead of how people were always 'entangled' with the physical world. And the result was that a common aim became one of seeking out 'entanglement examples' that show how we shouldn't think of ourselves as so special, so in control.

Yet important social trends can end up being downplayed when we work this way. Indeed, some of them have been exactly what this book has tried to reconcile with this agenda – how it may not be all that easy (or all that enjoyable) to be so entangled and how many may now be living some increasingly disentangled lives. Of course, there will be many exceptions. But it is, for example, common all around the world to see air-conditioning leading to the social avoidance of sweat, to see indoor gyms becoming the default exercise environment for many, for certain greenspaces to be controlled and paved over partly because that can make them easier for people to manage, and for certain facilities and conditions to be provided that allow people to go on as they usually do without giving much thought to environmental changes outside. It would be good to see more studies that acknowledge and explore the impacts of these broader trends, rather than searching for entanglement examples and then feeling reassured that they can still be found. Whilst we are busily doing that, it is quite possible that, without us necessarily noticing, the wider situation might be changing such that these spaces of experienced entanglement are increasingly, and to add a further environmental metaphor, endangered themselves.

Hunting for Moles

At this point, I should say something about the claims that can be reasonably associated with my results. This book is about how the various and varying environmental phenomena found outdoors may serve to unsettle everyday practices. In that sense, it was about looking for instances when practices had reached the point when they had become so entrenched that the idea of responding more fully to the situation outdoors was itself disruptive. Some might therefore say that, in the four projects that shape this book, I deliberately sought out a strange sort of person who embodied these attributes – a person who has been pulled away from the outdoors, who is so busy getting on with other tasks and objectives that, when they suddenly find themselves confronted by the situation outside, they would look up from their practices blinking and disorientated and unsurprisingly 'unsettled'. In this sense, those who I studied appear like moles who have been dragged out of their established preoccupations, either by the questions that I put to them in my interviews or by certain experiences in outdoor spaces.

And it is true that, within all four studies, I effectively set off on a sort of mole hunt – looking for situations and instances that allowed me to examine exactly this disorientation. But the point was always to extract insights 'from the edge' and to reflect on what they tell us about where social life was heading in specific contexts. Certainly, these processes were not evident in all domains of everyday life for these respondents. This book is about the times when we can all become a bit like moles, rather than those who consistently display this trait. It is quite possible, and indeed it was sometimes observed in my studies, that those who have been recruited into 'outdoor indifferent practices' in some domains of their lives were compensating for this in others. Some of the office workers, for example, spoke excitedly about weekend gardening or escaping to the countryside precisely because they felt they really needed it after a long week of labouring inside their offices. Meanwhile some of those who were troubled by the suggestion of an actively growing gardening spoke with passion about the pleasures of outdoor exercise. And, to be quite clear, there are many other groups (in both the United Kingdom and elsewhere) who live very different lives. Social trends don't just happen. They emerge gradually as some of us submit to them and others remind us through their words and their actions that we might not always want to.

Further Adventures in the Everyday Outdoors

On this more positive note, it is worth emphasising how the future of outdoor benefits in everyday life is far from decided. Whilst I deliberately trained my attention onto certain moles, or more accurately the times when we can all become mole-like, there are very many ways of examining (and potentially influencing) the processes involved. Other aspects of 'outdoor experience' could be usefully studied in this way too. Some further adventures on which researchers might embark in pursuit of the most positive social and environmental futures:

Weather and climate   Outdoor temperatures and local weather conditions currently do, and perhaps increasingly will, structure the extent to which it occurs to different groups of people to engage with outdoor environments in their everyday lives. With climate change predicted to present many of us with hotter and more variable weather, understanding how this complicates everyday practices is increasingly pressing. I have downplayed how this happens in this book because I wanted to focus on outdoor challenges as they emerged in terms of the specifics of the practices at hand. In other words, I wanted to be open to the various ways in which my practice carriers spoke of and responded to outdoor spaces, rather than to be too prescriptive in terms of settling on particular physical processes at the start. Nevertheless, there are many reasons for a fuller research focus on how hotter and more changeable climates will be responded to by relevant social practices as they move into the future.

This is partly because some of the processes of human cocooning seen in my projects are clearly spreading around the world, despite the continued variation in the climates that people live with and the variety of locally specific adaptation practices that still persist. London was the case study city that provided my focus, and London is a city with both mild and variable weather. Outdoor climates are very different elsewhere. Other cities are further along a route towards indoor living that seals the residents off from some much more extreme environmental challenges in terms of heat, cold, rain, humidity and others. Some of the ideas that I developed here could help in exploring and influencing their trajectories. After all, whether climate change at various time scales) is permitted to complicate everyday lives, and whether that proves a good thing, will depend on the extent to which the practices of which these lives are composed prove responsive to them.

Technology and time   Another trend that is set to play an increasingly important role in how and whether outdoor greenspace ends up punctuating everyday life relates to how certain technologies have already become, for some groups at least, a relatively constant companion. In this regard, whilst this book has been underpinned by the suggestion that, were people to spend time outside in greenspace, that might bring them various benefits, the truth of the matter is that getting them there might not always be enough. Part of their value, according to many of the studies discussed at the start, after all, relates to how greenspaces take us away from our immediate pressures – they allow us somehow to relax and then return to our lives refreshed. But we probably shouldn't assume that this will naturally happen as soon as people are there.

If, for example, many people are preoccupied with their technologies, or if they feel compelled to look at their phones (despite being in places that might benefit them, were they only to look up), then the challenge becomes more complicated. As we have seen in some of the case studies, there can be good reason to be reticent about clicking into the alternative (albeit restorative) states that greenspaces can encourage. If people feel they are busy, they may prefer to retain a more purposeful disposition. And that leaves us with questions about when people are more or less able to embrace the outdoors and how both technology and senses of time pressure can compel them to eschew potentially positive experiences. More, I think, could, for example, be done to compare weekend and weekday dispositions and to understand how smartphones and social media are acting to shape relationships with outdoor environments. If we want to understand the scope for providing greenspace benefits in the round, for example, perhaps the focus should be on local WiFi provision more than the most effective forms of landscape planning and design.

Dirt and disruption   Across the board in the four studies described in this book (though not always given prominence in the presented narratives because of the broader focus), I found a degree of anxiety about how the physicality of the outdoor environments involved would be experienced (Could I cope with this at lunchtime? What might this do to my running? What does the soil actually contain and can I handle that? Do I want to sit on certain surfaces at the festival and should I?). The question was one of whether doing so was appropriate or even enjoyable. The physicality of the experience is a topic that, as I mentioned at the start of the book, those interested in greenspace restoration have (rather like some of my research subjects) sometimes shied away from. But it could be crucial to examine in terms of the suggested need for human environments to be controlled and sanitised.

On the one hand, certain aspects of everyday life can be characterised, for some at least, by the increasingly efficient removal of what is taken to be 'dirty' – from our clothes, from our bodies, from the environments in which we spend our time. Yet, on the other hand, a collective recalibration of our various dirt relationships could be of benefit in a number of regards (going again beyond greenspace to resource consumption). The processes by which particular environmental features becomes situationally defined as 'dirt', 'earth', 'mud', 'soil' and so forth have implications in terms of what is deemed appropriate to do with them. How do developing practices accommodate these features?

Building Back Better

All these issues are not new. The point is rather that some of the ideas that I have developed here could help in understanding them. And, in that sense, a focus on disruption seems especially apposite right now. As I write, the world is reeling from the effects of the Covid-19 pandemic, which, in many ways, is encouraging a similar examination of the relationship between everyday life and outdoor spaces. The mole metaphor feels appropriate as many people have been shaken out of their usual practices and forced to reflect on their relationships with them. The hope, for some commentators, is that the result will be that societies 'build back better' by taking the opportunity to push practices towards 'new normals' that are preferable to the old.

Office work, for example, is no longer so tied to the office as we once assumed that it was and the idea of upholding professional dress codes may now feel faintly ridiculous if those involved are doing internet calls from the kitchen instead of inside meeting rooms. Gyms have been closed and treadmill runners have been forced outside in ways that make us wonder whether they will stay there after they have reopened. Being confined to homes has encouraged a wider revaluation of the benefits of domestic gardens, along with giving some people more time to nurture relationships with the plants found within them. Finally, not going outside will have implications for how and when people wash and, after protracted isolation, the escape of the festival might be especially craved. The pandemic underlines how practices are never really fixed. Some of the ways in which they have been studied in this book could help in encouraging them to take the best paths at points like these.

Intervention Ideas

Such statements demand that I say something about the role of social researchers in all this. Readers may be thinking that we can surely do something more than merely training our attention onto relevant processes and then worrying about them. Researchers, like myself, sometimes disappear at this point in the hope that those who read their reports will somehow magically translate their insights and observations into positive actions. So, whilst this book has put a spotlight on certain processes, could we also influence them? My answer is that how this is best done will depend on the situation at hand. I recognise that the reluctance to be more programmatic might be frustrating at this point. Surely, after having done all these interviews, I should be ready to weigh in with some overall suggestions about safeguarding and encouraging environmental benefits in everyday life? However, as I see it, to jump the gun in this way would be to discourage others from paying appropriate attention to the practices that interest them when each of them will be at least a little different. Instead, mine is more of a hopeful plea for sensitivity to the specifics of the context at hand and, in that respect (and looking forward to the next section of this concluding chapter), I would rather provide some suggestions about how to study them.

Yet, having said all that, one intriguing feature of some of my studies was how the act of taking part could itself serve as a trigger for change. One of the office workers, for example, quit his job soon after the last interview. He may have decided to take part in the project because he was already interested in thinking through his relationship with work. But it was also the case that the project encouraged him to reflect on his experience of outdoor spaces in a way that partly helped him to settle on a different career path. Similarly, a few of the indoor runners decided to relocate their running after having taken part in our interviews. They were keen on talking about how they 'really should' do that in our discussions and, though we were at pains to stop short of championing any particular running environment, taking the time to think about it proved a sufficient spur to action. In effect, they talked themselves off their treadmills.

We can't, of course, interview everyone in the hope that they will change their ways afterwards. And I don't want to fall into the trap of seeing the task as always being about getting people to think, and then act, differently. To do so would be to replicate some of the assumptions that the concepts on which I drew were designed partly to surpass. These include the idea that individuals are essentially the masters of their own destiny and our job, if we hope to promote positive social change, is ultimately one of identifying the right means of influencing their decisions. Indeed, despite the above observations about the indoor runners and the office workers, in the other two projects, the suggested targets were certain professional groups who structure these decisions – how garden designers could steer people towards certain positive garden relations and how festival organisers might recognise how they have a hand in making future attendees. In that sense, for these studies, the aim was to encourage these professional groups to recognise how they are often not responding to apparently fixed desires but actively creating the conditions that encourage sections of society to think and act in certain ways.

Wager Two – Studying 'Figures of Speech'

Having said all that about my first wager, I can now turn to my second. This was that, in more general terms, it is worth paying attention to how people speak of the practices that live through them. I argued at the start that, despite what others have sometimes assumed, we could learn a lot from taking an interest, by encouraging and exploring particular forms of talk, in the everyday lives of others. I now end by considering the various 'figures of speech' that were seen in my studies. This phrase fits nicely because it draws attention to how, under the circumstances of being asked to speak of certain aspects of their lives, people often reach for particular turns of phrase. And, as we have seen, this can happen partly because it makes life easier in the moment to sidestep further thought or discussion about particular topics. In other words, it encourages us to see the things that people say, both to us, to themselves, and to others, as responses to the situation at hand, rather than 'windows onto their world' in the way that researchers sometimes like to imagine them. But, beyond that, I like the idea of reversing the imagined direction of travel between talking and talkers because that encourages us to see those who we study partly as conduits for the words that our practices would have us say. The speech makes them as much as them calling forth particular phrases to convey the thoughts that were assumed to have come first. People can be 'figures of speech' in the sense that how they come to speak of their actions partly makes them who they are.

However, beyond that, I don't want to argue for a singular way of conceptualising this relationship. My four studies were indebted to a particular set of ideas that encouraged me to attend to certain features of how my interview respondents spoke with me about their lives. But my aim was also to notice the various ways in which these ideas could be seen in my four studies. So, in terms of overall conclusions about the three suggestions that I set out to explore at the start, my point is that, whilst all three applied, how they did definitely varied. The degree of exposure to the practice at hand, where the practice happens, what undertaking it physically involves: all of these, I think, served to colour the ways in which those involved spoke of their practices to me. I don't want to boil this all down into an overall evaluative response to each of my three suggestions. Instead, I would rather encourage studies that explore the extent to which other contexts are different or similar to my own. I did, after all, begin my second chapter with an argument against using certain experiences inside a boxing gym to make more general claims about the difficulties people are presumed to have in talking about their practices.

My final audience is therefore made up of those who are also hoping to understand the everyday lives of others by speaking with some of them. The power of practices partly stems from how they can become unremarkable to those involved. We might therefore anticipate various challenges in talking about them. And they are certainly there. But, if we hold our nerve, the result can be some absorbing exchanges in which processes that previously went unspoken become gradually apparent (both to the respondent and their researcher). This is not a matter of putting people under a stressful spotlight when we need to keep our respondents onside and we should definitely treat them well after they have been kind enough to volunteer at the start. The approach is rather about gently prodding at the practices that people carry as we gradually come to understand the relationships that they have with them based on the clues that can be seen in the things that they say.

Some clues to look out for in other studies:

Practice Dodging

Within all four of my studies, one revealing feature related to the extent to which those involved were happy to associate themselves with the practices they were carrying. Would those involved say they were competent office workers, devoted treadmill runners, skilled shoppers or keen showerers? And, if not, why not? The people involved were all of these things. But how they responded to the idea of being spoken of as such was another matter. So, what is happening when you admit to being such a person? One problem is that doing so can stray uncomfortably close to self-identification, irrespective of how much time is spent doing these things. And that may be why those who do not readily attach themselves to a practice can be quite good at faithfully reproducing it. It is partly by virtue of them not saying that they are particularly doing something, either to me in an interview or to themselves in everyday life, that their current felicity goes untroubled.

In all four of my case studies, those involved were ambivalent about being associated with the practice that I took as my focus and the reasons for this varied. For the runners, this process allowed them to keep doing something they thought was valuable. For the office workers, it allowed them to ignore certain humdrum features of their lives. For the garden shoppers, it allowed them to act in line with a practice they otherwise liked the idea of transcending. For the young showerers, the power of the practice was linked to how it was taken to be universal. The reasons will be different again for other practices. In the four case studies considered here, after all, I deliberately focused on practices that were either already widespread or that seemed to be increasingly common. And, if they are widespread, self-identification becomes more difficult (or less likely to seem worth doing).

This suggests some intriguing recruitment issues in terms of accessing those who don't think of themselves as particularly doing what interests us whilst at the same time doing quite a lot of it. One response to this challenge would be to focus on those who do it in particular ways, particular places or with certain levels of frequency. Doing so would allow the researcher to delay the issue of practice identification (or avoidance) until they are in a better position to explore how particular activities feature in people's lives. Either way, and regardless of whether we choose to see the people or the practices as the originators of this effect, how they handle the idea of being attached is instructive. So, when people say they are just having a shower (rather than showering), heading into the office (rather than embarking on some office work), going for run (rather than being a runner) or buying some things for the garden (rather than garden shopping), this tells us something about how and why they may want to duck the implications of a more wholehearted relationship with that practice. And this kind of practice dodging can have implications for what they do and the likelihood of them considering the idea of doing differently.

Following Our Leads

There are different ways of thinking about the 'leading question' in social research. In general, however, it's positioned as something that is tempting to pose because it might give us the answers that we want and, because of this, is best avoided. But how are we imagining our spoken exchanges when we take such a position? The implicit idea here is that the person with whom we are speaking will likely be compelled by the situation to go along with our leading questions, to present themselves in ways that fit with our misguided assumptions and don't really align with the actuality of their lives. To think this, however, is to buy into a certain vision of what people are – either that they are so eager to please that they will be happy to endorse our assumptions or, rather less charmingly, that they are so uninvested in the exercise of speaking with us that they'll just go with the flow and say whatever we want until the interview eventually ends.

We could also see the leading question as an opportunity to tap into the modes of thought tied to particular practices – as a route into exploring how we are all carried along by combinations of actions and thoughts in our everyday lives. In the four studies, for example, I asked people about the importance of keeping clean, about the processes through which they choose the most attractive garden plants, about the objectives that are woven into their current and future recreational running, and about the priorities that they had when thinking about their lunchtimes in the office. But, to me, this was not really 'putting words into their mouths'. Rather I think of this as an attempt to understand whether these ways of describing their relationship with the practice at hand were aligned with the combinations of thinking and doing that were already at large in their lives.

In all four of my projects, respondents contradicted themselves. The implication of how we commonly think about doing so is that this will likely feel like a shameful experience for people – as though they are suddenly discovering how disappointingly inconsistent they are in their approach to life. But we all are. My point, in other words, is that specific combinations of doing, thinking and speaking can, at certain times, take hold of us and, at others, an entirely different set can take their place. 'Leading questions' allow us to see how easily certain combinations creep in. After all, we are different people in different situations and, in this field of work particularly, part of the point is to attend to the pull and push processes involved instead of assuming that our objective is to coax out privately cherished ideas and opinions that (if we treat them right) people will eventually share with us. Whilst this should be carefully done, sometimes we should lead away (and examine how willingly they follow).

Analysis Avoidance

Unlike the leading question, the idea that people can talk about, or to sound more appropriately academic, reflexively examine, their situation is a cornerstone argument for doing interviews. As discussed at the start, the imagined encounter is one in which the researcher and their subject talk openly together and, through this process, come to understand how it really is to live in certain ways. Some of the authors introduced at the start were suspicious of this cheery vision of open conversation because their thinking was that much of how everyday life goes on is unconscious. As such, it is unlikely to be easy, and is certainly not enjoyable, to drag certain actions into the realm of active conscious evaluation through spoken exchange. But, like the idea of the leading question, I'd rather not see this in such stark terms. The truth of the matter is often somewhere between these two positions.

For the recreational runner, the challenge was perhaps most pronounced since there were reasons why active reflection on their running practice was best avoided. Similarly, for the fledgling gardener, this was a problem because it could lead to the dispiriting conclusion that they were finding it hard to embrace exactly the relationship with living plants that they otherwise wanted. Yet, for those at the festival and those in the office, this kind of analysis was eventually welcome, partly, I think, because the practices involved were not so associated with the idea of individual investment (and were instead seen as more socially regulated). It was taken to be difficult to deviate from perceived conventions in such cases and, partly because of that, it was less risky to consider the practice.

My point here is that, rather than engage in too much abstract debate about whether scrutinising our practices though interview exchange is possible, we might pay more attention to occasions of analysis avoidance, namely the circumstances under which people pull back from this activity. In order to understand how patterns of everyday life become entrenched, we can learn from how and when such avoidance happens. And again, these reasons will vary. In the garden, inside the office, on the treadmill or at the festival: the ways in which respondents managed this aspect of our discussions differed. And they will likely be different again elsewhere. Analysis avoidance could provide a useful focus for investigations of how life goes on in all sorts of places, since each will have their own revealing quirks in terms of how forms of practice talk are embraced or eschewed.

Changing Tack

Some of the above characterisations make it seem like we are asking people to embark upon a journey of (sometimes enjoyable, sometimes unsettling) self-discovery when we encourage them to speak about activities they may otherwise do as a matter of course. And it is probably true that social researchers and those from whom they hope to learn in their studies have a rather different relationship with this kind of analysis. I do my job partly because I'm interested how and why particular patterns of everyday living become entrenched. And that certainly did, in all of the four cases, sometimes mean posing questions that felt out of step with the natural back-and-forth of appropriate and acceptable exchange. It sometimes also prompted respondents to change the conversational tack. Without always noticing this redirection, in doing so, they shifted the focus from an intimate examination of their existing personal practice to a discussion of either how things went for some wider collective or another or how things would ideally go for them in principle.

By now my position on such moments is predictable. I think they are often quite revealing in ways that make them worth enduring a few seconds of situational discomfort (that is likely much more pronounced for us, since we are professionally invested in the exercise, than them). When those with whom we talk reframe the question, the times when they dodge certain forms of spoken examination, and the strategies they deploy to change the conversational tack: these are all features of an interview that tell us something. They are much more than mere bumps on the road in an otherwise freely flowing exchange (as if things ever really worked like that). Indeed, looking out for these reactions (and thinking about what triggered them) takes us, I think, closer to the lived realities of taking part in social practices when what is sayable, thinkable and doable are often interlinked.

Endnote

1 This is, of course, only one way of imagining how things will play out. Future societies could equally be inspired to take action by the nature programmes they watch on
TVs and laptops indoors. News reports of catastrophic climate change impacts might
more effectively trigger a collective demand for systemic overhaul. There are also the
grassroots movements that are pushing hard for positive change. The point here is more
about examining existing trends and working through their implications.

Reference

Vivid Economics. (2017). *Natural Capital Accounts for Public Green Space in London:
Report Prepared for Greater London Authority, National Trust and Heritage Lottery Fund.*
London: Vivid Economics.

Index

The Unsettling Outdoors: Environmental Estrangement in Everyday Life, First Edition. Russell Hitchings.
© 2021 Royal Geographical Society (with the Institute of British Geographers). Published 2021 by John Wiley & Sons Ltd.